ntinued on back)

D1271153

Activation Spectrometry
in Chemical Analysis

CHEMICAL ANALYSIS

A SERIES OF MONOGRAPHS ON
ANALYTICAL CHEMISTRY AND ITS APPLICATIONS

Editor
J. D. WINEFORDNER
Editor Emeritus: **I. M. KOLTHOFF**

VOLUME 119

A WILEY-INTERSCIENCE PUBLICATION

JOHN WILEY & SONS

New York / Chichester / Brisbane / Toronto / Singapore

Activation Spectrometry
in Chemical Analysis

SUSAN J. PARRY

Imperial College Reactor Centre
Imperial College of Science, Technology and Medicine
University of London

A WILEY-INTERSCIENCE PUBLICATION

JOHN WILEY & SONS

New York / Chichester / Brisbane / Toronto / Singapore

Copyright © 1991 by John Wiley & Sons, Inc.

All rights reserved. Published simultaneously in Canada.

Library of Congress Cataloging in Publication Data:

Parry. Susan J.
    Activation spectrometry in chemical analysis / Susan J. Parry.
       p. cm. — (Chemical analysis: v. 119. ISSN 0069-2883)
    "A Wiley–Interscience publication."
    Includes bibliographical references.
    ISBN 0-471-63844-7
    1. Nuclear activation analysis. I. Title. II. Series.
QD606.P37   1991
543'.0882—dc20   91-12496
                                                          CIP

Printed in the United States of America

10 9 8 7 6 5 4 3 2 1

To Maurice Kerridge
and my family

PREFACE

Activation spectrometry, which forms the basis of most routine work in the field of activation analysis, is the combination of neutron activation and gamma ray spectrometry. This book is an introduction to the technique of activation analysis as it is currently applied to analytical problems, and therefore concerns mainly activation spectrometry but includes other activation methods. It is my view that activation analysis has finally been accepted as a routine method of trace element analysis. It is possible to provide industry with a very successful, rapid and reliable service which is competitive with other techniques.

It is therefore increasingly important to inform colleagues in other disciplines about what can be achieved with the method. My own experience of teaching postgraduate students from a wide range of disciplines has shown me how important it is to educate potential users so that they can use activation analysis in the most effective way; even if they do not carry out the work themselves. I have written this book for chemists, geologists, biologists, physicists, environmentalists, and archaeologists; in fact anyone who wishes to find out about the technique and how it may be applied in her or his field.

The description of any analytical method falls naturally into three sections: principles, techniques and applications. My text follows the same pattern, after a short introduction to activation spectrometry in the first chapter. Chapters 2 to 6 give a simple outline of the principles behind the technique, without introducing the detailed physics which has been described so well in other publications. Further information on the topic may be found in the references which are given at the end of each chapter. Potential users may be more interested in the technique itself and so in Chapters 7 to 10 I have described sampling and preparation of material, how to make standards and which reference materials to use. Chapters 11 and 12 are designed to help new users to choose the best analytical conditions for their particular determination. Finally I have devoted Chapters 13 to 16 to biomedical, environmental, geological and industrial applications; reference to other areas such as archaeology are also found under these headings where relevant. These chapters are not meant to provide a comprehensive review of all the applications of the method but rather a guide to new users to the type of routine work which activation analysis has been applied to.

To summarize, I have written this book for undergraduates and postgraduates, in universities, research institutes, government agencies or industry, as a guide to the method as it is applied on a day-to-day basis for routine analysis.

I am grateful to all the staff and students at Imperial College Reactor Centre, for their support and encouragement while I was writing this book. In particular I would like to thank Brian Bennett, Roger Benzing, Desmond MacMahon and Margaret Minski who also made contributions in the form of expertise, advice and unpublished data. Finally my warm thanks go to Joyce Vine who prepared many of the tables and figures in the manuscript.

SUSAN J. PARRY

January 1991
Imperial College Reactor Centre
Ascot, Berkshire, United Kingdom

CONTENTS

Activation Spectrometry
in Chemical Analysis

CHAPTER

1

WHAT IS ACTIVATION SPECTROMETRY?

Whenever an analytical technique is being evaluated for use on a particular problem, there are a number of questions which will be asked about its performance. The user will want to know how it works and what it is capable of in terms of accuracy, precision and sensitivity; whether it is affected by matrix or concentration; how long an analysis will take and whether large numbers of samples can be handled; and finally, how much it will cost? In this introductory chapter we will consider the answers to some of these preliminary questions in relation to activation spectrometry.

What is the definition of activation spectrometry? The term activation spectrometry is defined as neutron activation analysis with gamma ray spectrometry. In activation analysis generally neutrons are the most common means of activation although charged particles will be implemented for the lighter elements. The term neutron activation analysis is normally assumed to mean analysis by gamma ray spectrometry unless an alternative is specified, such as in the case of delayed neutron analysis. Activation spectrometry then refers to the activation of an element with neutrons and its subsequent determination using gamma ray spectrometry.

What is involved? The sample to be analyzed is placed in a neutron flux, usually in a nuclear reactor. It is irradiated to activate the elements of interest and transferred to a gamma ray detector. The gamma rays emitted by the radionuclides are counted for a measured time period. Gamma ray spectrometry is used to identify and measure up to about 40 elements simultaneously.

What form does the sample have to take? In general the form of the sample is not important. It may be a solid, liquid, powder, even a gas. However the shape and size of the sample will affect the specific activity induced on irradiation and counting and so it is important to be aware of the problems that may occur. The range of possible sample sizes is a valuable feature of the technique and weights from micrograms to hundreds of grams have been analyzed by this method.

1

Table 1.1. Determination limits for single element standards: a comparison of neutron activation with other analytical techniques

Element	FAAS[a] (mg dm^{-3})	ICP–AES[b] (mg dm^{-3})	ICP–MS[c] (mg dm^{-3})	INAA[d] (mg kg^{-1})
Al	0.06	0.046	0.00097	0.001
Sb	0.12	0.064	0.00028	0.00005
As	0.003	0.106	0.0028	0.00001
Ba	0.06	0.0026	0.0002	0.001
Bi	0.12	0.272	0.00001	0.010
B	2.1	0.0096	0.00038	–
Cd	0.003	0.0050	0.00057	0.0005
Ca	0.003	0.020	0.010	0.10
C	–	0.35	–	–
Ce	–	0.104	0.0001	0.0001
Cs	0.06	83.0	0.00009	0.0001
Cr	0.009	0.014	0.00011	0.003
Co	0.015	0.012	0.00034	0.00001
Cu	0.009	0.0108	0.00023	0.00001
Dy	0.15	0.054	0.0001	0.000 001
Er	0.12	0.036	0.00006	0.0001
Eu	0.09	0.0054	0.00006	0.00001
Gd	6.0	0.050	0.0001	0.0001
Ga	0.21	0.092	0.00098	0.00005
Ge	0.3	0.096	0.00096	0.010
Au	0.03	0.034	0.00011	0.000 0001
Hf	6.0	0.030	0.00002	0.0001
Ho	0.18	0.0114	0.00004	0.0002
In	0.12	0.126	0.00026	0.0001
Ir	1.5	0.056	0.00008	0.000 001
Fe	0.015	0.0124	0.001	0.10
La	6.0	0.020	0.00008	0.00001
Pb	0.06	0.084	0.0003	1.0
Li	0.003	0.045	0.00029	–
Lu	0.9	0.0020	0.00005	0.00001
Mg	0.0006	0.060	0.00094	0.010
Mn	0.006	0.0028	0.00029	0.00001
Hg	0.51	0.050	0.00023	0.0001
Mo	0.06	0.0158	0.0007	0.001
Nd	3.0	0.0150	0.0002	0.0005
Ni	0.015	0.030	0.004	0.010
Nb	6.0	0.072	0.00003	0.001

Table 1.1. Continued

Element	FAAS[a] (mg dm^{-3})	ICP–AES[b] (mg dm^{-3})	ICP–MS[c] (mg dm^{-3})	INAA[d] (mg kg^{-1})
Os	0.3	0.0007	0.00044	0.001
Pd	0.045	0.088	0.0041	0.00005
P	120.0	0.152	0.032	0.001
Pt	0.21	0.11	0.00041	0.0005
K	0.003	12.0	0.004	0.0002
Pr	15.0	0.094	0.00009	0.00001
Re	2.4	0.012	0.004	0.00001
Rh	0.015	0.088	0.002	0.0001
Rb	0.015	75	0.00036	0.0001
Ru	0.9	0.060	0.00044	0.001
Sm	3.0	0.086	0.0002	0.000 005
Sc	0.15	0.0030	0.00006	0.00001
Se	0.6	0.15	0.00028	0.005
Si	0.45	0.024	0.033	0.010
Ag	0.006	0.00032	0.00032	0.001
Na	0.001	0.058	0.003	0.0001
Sr	0.015	0.00084	0.0001	0.010
Ta	4.5	0.050	0.00001	0.00005
Te	0.15	0.082	0.003	0.010
Tb	3.0	0.056	0.00003	0.0001
Tl	0.06	0.080	0.00003	0.010
Th	–	0.166	0.00001	0.0001
Tm	0.3	0.0104	0.00001	0.00001
Sn	0.3	0.09	0.00043	0.001
Ti	0.21	0.0076	0.0024	0.003
W	3.6	0.060	0.00001	0.00001
U	90.0	0.500	0.00002	0.00001
V	0.15	0.015	0.00016	0.00001
Yb	0.12	0.0036	0.00006	0.00001
Y	0.6	0.0070	0.0001	0.0001
Zn	0.003	0.0036	0.00059	0.010
Zr	4.5	0.0142	0.00011	0.100

Sources: [a] Flame atomic absorption spectrometry (Potts, 1987), [b] Inductively coupled plasma atomic emission spectrometry (Potts, 1987), [c] Inductively coupled plasma mass spectrometry (Jarvis, unpublished; Jarvis and Williams, 1989; Potts, 1987), [d] Instrumental thermal neutron activation analysis (Revel, 1987)

What is the precision of the method?   The technique is noted for its precision and that is why it is frequently used as the reference technique for the development of reference materials. The main limitation to precision is due to the counting statistics in the gamma ray spectrometry and usually it is possible to improve precision by long counting times. The exceptions to this are elements which activate to give short-lived radionuclides.

What is the sensitivity of the method?   The technique is generally used for trace element analysis for multielements. The detection limit is at the microgram level for many elements and down to nanogram quantities for some. However the detection limit will depend on the other elements present in the matrix. Table 1.1 lists the determination limits for elements by neutron activation analysis under ideal conditions, assuming a thermal neutron flux of 10^{18} n m^{-2} s^{-1} for three days (Revel, 1987). The determination limits are compared with those obtained with the other main multielement methods of analysis (Potts, 1987; Jarvis and Williams, 1989).

What is the effect of concentration?   The method can be used to determine elements at a wide range of concentrations since the calibration "curve" is actually a straight line. In theory, the method can been used to measure elements over concentration ranges from nanograms to grams. However, if the element is present in too high a concentration there is a possibility of self-shielding, which may reduce the specific activity of the radionuclide produced.

What is the effect of the matrix?   Elements that give intense gamma ray activity on irradiation will produce background activity and possibly interfering gamma rays. Some elements do not activate on irradiation with thermal neutrons to produce gamma rays and so, for example, carbon, hydrogen, nitrogen, and oxygen will not affect the analysis. On the other hand aluminum, sodium, and chlorine activate very well and will cause serious interferences which must be overcome.

What is the turnaround time?   The time for an analysis depends on the elements to be measured. If the element activates to give a short-lived nuclide, it may be irradiated and counted in a few minutes. On the other hand, some elements may be analyzed using an irradiation of several weeks and left for months before counting for many hours. The timings are relatively flexible for the longer-lived radionuclides and the analyses can take as long as there is time available.

Table 1.2. Determination limits for silicate rock: a comparison of neutron activation with other analytical techniques

Element	FAAS[a] (mg kg⁻¹)	ICP–AES[b] (mg kg⁻¹)	ICP–MS[c] (mg kg⁻¹)	INAA[d] (mg kg⁻¹)
Al	12	9	0.485	100
Sb	5	20	0.14	0.1
As	60	15	1.4	1
Ba	10	1	0.1	20
Bi	5	20	0.005	–
B	90	2	0.19	1
Cd	2	5	0.285	–
Ca	0.6	4	5	200
C	–	70	–	–
Ce	–	35	0.05	3
Cs	12	16 600	0.045	0.2
Cr	10	3	0.055	0.5
Co	5	5	0.17	0.1
Cu	5	5	0.115	–
Dy	30	15	0.05	0.2
Er	21	15	0.03	–
Eu	18	2	0.03	0.5
Gd	1 200	5	0.06	3.9
Ga	42	16	0.49	20
Ge	60	40	0.48	–
Au	0.1	10	0.055	0.005
Hf	1 200	6	0.01	0.2
Ho	36	15	0.02	0.1
In	24	25	0.13	0.2
Ir	300	11	0.04	0.005
Fe	5	2.5	0.5	50
La	1 200	5	0.04	0.1
Pb	10	20	0.15	–
Li	3	3	0.15	–
Lu	180	0.4	0.025	0.05
Mg	0.12	12	0.47	–
Mn	5	15	0.15	100
Hg	102	10	0.12	–
Mo	3	10	0.35	2
Nd	3	25	0.10	5
Ni	5	5	2	50
Nb	90	5	0.02	–
Os	60	0.14	0.22	10

Table 1.2. Continued

Element	FAAS[a] (mg kg⁻¹)	ICP–AES[b] (mg kg⁻¹)	ICP–MS[c] (mg kg⁻¹)	INAA[d] (mg kg⁻¹)
Pd	9	18	0.21	10
P	24 000	30	16	–
Pt	42	22	0.21	–
K	0.6	2 400	2	–
Pr	3 000	40	0.05	–
Re	480	2.4	2	1
Rh	3	18	1	–
Rb	1	15 000	0.18	0.1
Ru	180	12	0.22	–
Sm	600	15	0.1	0.1
Sc	10	2	0.03	0.05
Se	120	50	0.14	0.5
Si	90	5	16.5	–
Ag	2	5	0.16	2
Na	0.2	12	1.5	10
Sr	1	5	0.05	100
Ta	900	10	0.01	0.03
Te	30	16	1.5	–
Tb	600	80	0.02	0.1
Tl	12	16	0.02	–
Th	–	–	0.01	0.2
Tm	60	2	0.01	0.34
Sn	20	6	0.22	–
Ti	42	1.5	1.2	100
W	90	50	0.01	1
U	90	80	0.01	0.1
V	1	5	0.08	1
Yb	24	1	0.03	0.2
Y	50	5	0.05	–
Zn	0.8	5	0.30	10
Zr	90	6	0.06	100

Sources: [a] Flame atomic absorption spectrometry (Potts, 1987), [b] Inductively coupled plasma atomic emission spectrometry (Potts, 1987), [c] Inductively coupled plasma mass spectrometry, 3 sigma assuming dilution × 500 (Jarvis, 1988 and unpublished data), [d] Instrumental thermal neutron activation analysis (Potts, 1987; Parry, unpublished data)

How much does it cost? The cost depends on the detection limits required and which elements are to be measured in what matrix. Typical commercial charges are competitive with other techniques which provide an equivalent range of elements and detection limits. The main potential cost is that of running a nuclear reactor. Normally the reactor will have been built for some other purpose, such as nuclear research and training, materials testing and isotope production. Neutron activation analysis is a useful by-product of the operation of the reactor and therefore the cost of irradiations can usually be kept to an affordable level.

What is it used for? Activation spectrometry has been applied to a wide range of sample types including biomedical samples, for example blood, tissue, hair, teeth and bones; environmental samples such as air filters, water, plant and vegetation; geological samples including rocks, minerals and ores; and a variety of industrial applications with carbon and boron matrices, organics, metals and alloys, glasses and ceramics.

How does it compare to other analytical techniques? In general activation spectrometry is one of the most sensitive methods for multielement analysis, with inductively coupled plasma emission spectrometry and inductively coupled plasma mass spectrometry. Activation spectrometry has the advantage that solid samples can be analyzed. A comparison of sensitivities for the various methods must be made with "real" samples and Table 1.2 lists the detection limits for elements in geological material for each of the techniques. The compilation contains information from a variety of unpublished sources and publications by Potts (1987) and Jarvis (1988).

REFERENCES

Jarvis, K. E. (1988), "Inductively coupled plasma mass spectrometry: a new technique for the rapid or ultra-trace level determination of the rare earth elements in geological materials," *Chem. Geol.*, **68**, 31–39.

Jarvis, K. E. and J. G. Williams (1989), "The analysis of geological samples by slurry nebulisation inductively coupled plasma mass spectrometry (ICP–MS)," *Chem. Geol.*, **77**, 56–63.

Potts, P. J. (1987), *A Handbook of Silicate Rock Analysis*, Blackie, Glasgow.

Revel, G. (1987), "Present and future prospects for neutron activation analysis compared to other methods available," in IAEA, *Comparison of Nuclear Analytical Methods with Competitive Methods*, IAEA–TECDOC–435, International Atomic Energy Agency, Vienna, pp. 147–162.

CHAPTER

2

NEUTRON ACTIVATION

Neutron activation is the irradiation of a nucleus with neutrons to produce a radioactive species, usually referred to as the radionuclide. The number of radionuclides produced will depend on the number of target nuclei, the number of neutrons and on the factor called the cross section which defines the probability of activation occurring. If the activation product is radioactive, it will decay with a characteristic half-life. Consequently the growth of activity during irradiation will depend on the half-life of the product. The energy of the neutrons which are bombarding the nucleus will dictate the type of interaction that occurs and consequently the nature of the activation product. Therefore, if the nucleus is irradiated in a neutron flux of both slow and fast neutrons, there may be more than one activation product. Similarly, interferences may occur as the result of the same radionuclide being produced by the activation of different target nuclei.

NUCLEAR REACTIONS

Neutrons can exhibit a wide range of energies. In a thermal nuclear reactor, for example, the neutrons are moderated and the majority are thermal or slow neutrons with an average energy of about 0.025 eV. A neutron generator is designed to produce fast neutrons with an energy of 14 MeV. The region between thermal and fast neutrons is called the "epithermal region," extending from 0.5 eV to 1 MeV. The lower cutoff at 0.5 eV is defined by the energy below which neutrons will not pass through 1 mm thick cadmium. It is therefore sometimes referred to as the cadmium cutoff.

A neutron is absorbed by the target nucleus to produce a highly energetic state of the resulting nucleus containing an additional neutron, and the excess energy is immediately lost, usually by emission of a gamma ray, a proton or an alpha particle. The energy of the neutron will affect the nature of the nuclear reaction which occurs and consequently the activation product. The main reaction occurring with thermal neutrons is the so called (n,γ) reaction. In this case the highly energetic level of the product nucleus is de-excited by emission of a gamma ray which is called a prompt gamma, since it is emitted immediately after activation:

$$^{27}_{13}\text{Al} + \text{n} \rightarrow {}^{28}_{13}\text{Al} + \gamma$$

Fast neutrons induce different reactions. The absorption occurs with either the ejection of a proton, in the case of an (n,p) reaction for example:

$$^{31}_{15}\text{P} + \text{n} \rightarrow {}^{31}_{14}\text{Si} + \text{p}$$

or the production of an alpha particle, as in the case of an (n,α) reaction:

$$^{23}_{11}\text{Na} + \text{n} \rightarrow {}^{20}_{9}\text{F} + \alpha$$

The gamma rays, protons and alpha particles produced during these reactions are all emitted spontaneously and therefore they are only detected if they are monitored during the activation process.

It is possible to see all three common reactions: (n,γ), (n,p) and (n,α), occurring with one target nucleus, for example in the case of ^{23}Na where not only the (n,α) reaction shown above occurs, but also the (n,γ) and the (n,p) reactions:

$$^{23}_{11}\text{Na} + \text{n} \rightarrow {}^{24}_{11}\text{Na} + \gamma$$

$$^{23}_{11}\text{Na} + \text{n} \rightarrow {}^{23}_{10}\text{Ne} + \text{p}$$

So in a neutron flux consisting of slow and fast neutrons the activation of ^{23}Na will result in the production of ^{24}Na, ^{20}F and ^{23}Ne. These activation products are all radioactive but the product of a neutron-induced reaction may be a stable, naturally occurring isotope of the particular element.

Whether the product is stable or not will depend on the ratio of neutrons to protons in the nucleus. An element may have several isotopes with a range of mass numbers. In stable isotopes, the ratio of neutrons to protons is close to 1.0 for light elements and increases to about 1.5 for heavy elements. If the ratio is outside the ratio band for stable isotopes a nucleus will be unstable and it will revert to a stable state by radioactive decay. For example ^{23}Na has 11 protons and 12 neutrons and is stable, while ^{22}Na which has only 11 neutrons and ^{24}Na which has 13 neutrons are both unstable and are radioactive. A nucleus may therefore be made radioactive by absorption of an additional neutron into the nucleus but it is, however, quite possible that addition of one neutron to a stable nucleus will produce a second stable product nucleus. For example in the case of magnesium there are three stable isotopes: ^{24}Mg, ^{25}Mg and ^{26}Mg. The following nuclear reactions will occur with thermal neutrons: $^{24}Mg(n,\gamma)^{25}Mg$, $^{25}Mg(n,\gamma)^{26}Mg$ and $^{26}Mg(n,\gamma)^{27}Mg$. The only radioactive product is ^{27}Mg, the other two products are stable isotopes of magnesium.

ACTIVATION WITH NEUTRONS

In a neutron-induced reaction, the growth of the product is dependent on the size of the neutron flux. The larger the neutron flux, the greater the rate at which interactions occur:

$$\text{Activation rate} \propto \text{neutron flux } (\phi)$$

The activation rate is also directly proportional to the number of target nuclei present:

$$\text{Activation rate} \propto \text{number of nuclei present } (N)$$

The number of target nuclei present will depend on the isotopic abundance of the particular isotope of interest. For example, aluminum is composed entirely of stable ^{27}Al and so all the target nuclei will be the same. Avogadro's number (N_A) represents the total number of atoms in the atomic weight (A) of any element. Therefore Avogadro's number, divided by the atomic weight gives the total number of atoms per gram:

$$N = N_A/A$$

and for a mass w, of the element, the total number of target nuclei will be :

$$N = w\, N_A/A$$

However, there may be more than one isotope of an element, such as in the case of calcium where there are six stable isotopes: ^{40}Ca, ^{42}Ca, ^{43}Ca, ^{44}Ca, ^{46}Ca and ^{48}Ca. As an example, ^{48}Ca is only present as 0.185% of the total. In such cases the number of target nuclei must be corrected for the isotopic abundance (θ):

$$N = w\, N_A\, \theta/A$$

The number of target nuclei is therefore proportional to the mass of

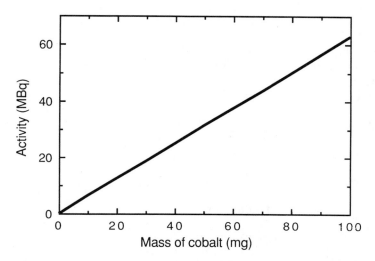

Figure 2.1. Standardization curve for cobalt, irradiated in a thermal neutron flux of 10^{16} n m^{-2} s^{-1} for 10 d, showing the linear relationship between the mass of the element and the induced activity.

element present and since the growth of the activation product is proportional to the number of target nuclei it follows that the activation rate is proportional to the mass of the element:

$$\text{Activation rate} \propto \text{mass of element } (w)$$

It is therefore possible to deduce the mass of the element present from the induced activity. This forms the basis of the neutron activation analysis technique. If the neutron flux remains constant then the "calibration curve" for an element can be determined by plotting the induced activity against the mass of the element. The simple relationship is shown in Figure 2.1 for ^{60}Co using the data from the compilation of calculated activities published by Bujdoso et al. (1973). The activity is given for a 10 d irradiation in a thermal neutron flux of 10^{16} n m^{-2} s^{-1}. The slope is equivalent to the specific activity and goes through zero.

CROSS SECTION

The relationship between activation rate, the number of target nuclei and the neutron flux is expressed by the term "cross section" (σ). The cross section is simply a physical constant:

$$\text{Activation rate} = \sigma \phi N$$

N is the number of target nuclei, in atoms
ϕ is the neutron flux, in neutrons $m^{-2} s^{-1}$
σ is the cross section, in m^2
activation rate is in events s^{-1}

Substituting:

$$N = w N_A \theta / A$$

into the expression for activation rate, it becomes:

$$\text{Activation rate} = \sigma \phi w N_A \theta / A$$

Cross sections are usually expressed in barns which are 10^{-28} m^2. As a rough guide, a target nucleus with a cross section in the order of barns will activate well but a cross section in millibarns indicates poor activation. It is important to remember that each stable isotope of the same element will have a different cross section. Consequently, one isotope may have a high cross section and become very active while another isotope of the same element may have a small cross section and be activated to a much smaller extent. It is therefore important to consider the cross sections when deciding which target nuclide to use in activation analysis.

The neutron cross section for a particular nucleus will depend on the energy of the neutron. Many nuclei, particularly of low atomic number absorb thermal neutrons with cross sections which decrease linearly with increasing velocity of the neutron (known as $1/v$ absorbers). It is usual to refer to thermal cross sections for the absorption of neutrons with an average velocity of 2200 m s^{-1}. Tables of cross sections are available for activation with neutrons. In the tables the cross sections may be expressed in different forms and the total cross section given for a particular target will be composed of a number of partial cross sections, dependent on the activation process, including (n,γ), (n,p) and (n,α) reactions. However for most thermal neutron activation the main process is the (n,γ) reaction involving the neutron radiative capture cross section (σ_γ).

Not all target nuclei are $1/v$ absorbers and there are many examples of nuclei which preferentially absorb epithermal neutrons. At these higher energies the neutron cross section is referred to as the resonance integral and the radiative capture resonance integral (I_γ) is used. The values for capture cross sections and resonance integrals are given by Mughabghab et al. (1984) and some typical examples are shown in Table 2.1. It can be seen from the cross section values that the lighter elements have

Table 2.1. Capture cross sections (σ_γ) and resonance integrals (I_γ) (in barns) for some typical activation targets

Target	σ_γ	I_γ	Target	σ_γ	I_γ
^{23}Na	0.4	0.31	^{63}Cu	4.5	5.0
^{26}Mg	0.038	0.026	^{75}As	4.3	61
^{27}Al	0.23	0.17	^{81}Br	2.4	60
^{37}Cl	0.43	0.30	^{109}Ag	91	1400
^{48}Ca	1.09	0.89	^{139}La	8.9	11.8
^{51}V	4.9	2.7	^{152}Sm	206	2970
^{50}Cr	15.9	7.8	^{151}Eu	9200	3300
^{55}Mn	13.3	14.0	^{186}W	37.9	485
^{58}Fe	1.28	1.7	^{191}Ir	954	3500
^{59}Co	37.2	74	^{197}Au	98.7	1550

Source: Mughabghab et al., 1984

thermal cross sections and resonance integrals in the same order. They are the $1/v$ absorbers. On the other hand the isotopes ^{109}Ag, ^{152}Sm and ^{197}Au have very large resonance integrals compared to the thermal cross sections, indicating that there are strong resonances in the region above the cadmium cutoff energy. In these cases it is important to include the resonance integral term in the calculation of the activation rate:

$$\text{Activation rate} = \sigma_\gamma\, \phi_{\text{th}}\, N + I_\gamma\, \phi_{\text{epi}}\, N$$

ϕ_{th} is the thermal neutron flux
ϕ_{epi} is the epithermal neutron flux

DECAY RATE

If the product nuclide in a neutron-induced reaction is stable the number of nuclei produced is easily calculated from the activation equation by multiplying by the length of irradiation, t:

$$\text{Activation rate} = \sigma\, \phi\, N$$

$$\text{Number of nuclei} = \sigma\, \phi\, N\, t$$

However, if the product nuclide is radioactive it will have a decay rate which must be taken into account. The radionuclide produced will decay with a characteristic half-life. If there are N^* radioactive nuclei, the rate of decay of the nuclei is proportional to N^*:

$$\text{Decay rate, } dN^*/dt \propto -N^*$$

$$= -\lambda N^*$$

where λ is the decay constant, which has a characteristic value for each radionuclide. If the equation is integrated between the limits N_0^* at time zero, and N^* remaining at time t:

$$N^* = N_0^* \exp(-\lambda t)$$

It is from the above expression that the term half-life is derived since, for if the time for half the nuclei to decay is defined as $T_{1/2}$:

$$N_0^*/2 = N_0^* \exp(-\lambda T_{1/2})$$

$$T_{1/2} = \ln2/\lambda = 0.693/\lambda$$

A semilogarithmic plot of the decay rate against time will give a straight line graph with a slope of $-\lambda$. Figure 2.2 shows the plot that would be obtained for ^{56}Mn which has a half-life of 2.58 h. The half-life of the radionuclide can be read directly from the time taken for the decay rate to be reduced by a half. A table of the half-lives, taken from Browne

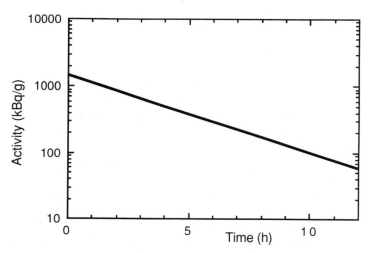

Figure 2.2. Decay curve for ^{56}Mn ($T_{1/2} = 2.58$ h), showing the semilogarithmic relationship between length of decay and remaining activity.

Table 2.2. Half-life values for typical activation products

Product	Half-life	Product	Half-life
^{24}Na	14.57 h	^{64}Cu	12.70 h
^{27}Mg	9.46 min	^{76}As	26.3 h
^{28}Al	2.24 min	^{82}Br	35.3 h
^{38}Cl	37.2 min	^{110}Ag	24.6 s
^{49}Ca	8.72 min	^{140}La	40.3 h
^{52}V	3.75 min	^{153}Sm	46.7 h
^{51}Cr	27.70 d	^{152}Eu	13.3 y
^{56}Mn	2.578 h	^{187}W	23.9 h
^{59}Fe	44.5 d	^{192}Ir	73.8 d
^{60}Co	5.27 y	^{198}Au	2.69 d

Source: *Table of Radioactive Isotopes*, Browne and Firestone, Copyright © 1986, reprinted by permission of John Wiley & Sons, Inc.

and Firestone (1986) for some radionuclides commonly measured by neutron activation analysis are given in Table 2.2. Because the half-life is characteristic for a particular radionuclide it can be used to identify an unknown species or confirm the identity of the radionuclide being measured.

INDUCED ACTIVITY

If the activation product is radioactive and decays with its characteristic half-life, the radionuclide is being produced at the rate described by the activation equation and decaying with the characteristic half-life. Consequently the growth of the activity is governed by the difference between them:

$$\text{Production rate} = \text{activation rate} - \text{decay rate}$$

$$dN^*/dt = \sigma \phi N - \lambda N^*$$

$$N^* = \sigma \phi N (1 - \exp(-\lambda t))/\lambda$$

The activity or disintegration rate (A_0), at the end of the irradiation time t, is then:

$$A_0 = \lambda N^* = \sigma \phi N (1 - \exp(-\lambda t))$$

Consequently the growth of the induced activity with time is controlled by the half-life of the activation product. This is demonstrated in Figure 2.3, where the growth curve for ^{49}Ca is plotted with data taken from Bujdoso et al. (1973). It can be seen that the majority of the activity is produced during the first two half-lives. When the irradiation time is very long the expression for activity becomes close to the maximum possible activity for a particular neutron flux, called the saturation activity (A_S):

$$A_S = \sigma \phi N$$

The saturation activity is independent of the half-life of the activation product and depends only on the value of the neutron flux and neutron cross section. A plot of activity induced in ^{64}Cu for different neutron fluxes in Figure 2.4 shows the growth of activity to saturation and how the saturation activity increases with neutron flux (Bujdoso et al., 1973). Unless the activation product is relatively short-lived it is not convenient to allow the growth curve to reach saturation. The usual form of the equation for activity at the end of an irradiation for a time t is:

$$A_0 = \sigma \phi N (1 - \exp(-\lambda t))$$

It is possible to calculate the induced specific activity for a particular

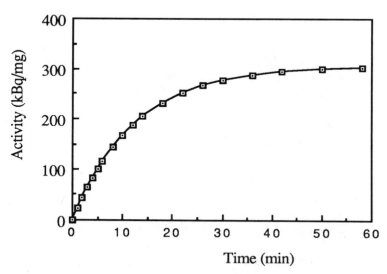

Figure 2.3. Activation curve for ^{49}Ca ($T_{1/2} = 8.72$ min), showing how saturation activity is approached after several half-lives.

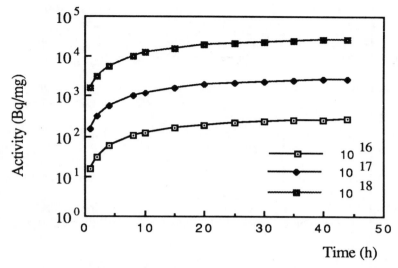

Figure 2.4. Activation curves for ^{64}Cu ($T_{1/2}$ = 12.70 h), showing the increase in saturation activity with increasing neutron flux, from 10^{16} to 10^{18} n m^{-2} s^{-1}.

length of irradiation, knowing the nuclear constants for the nuclide of interest and the neutron flux:

$$A_0 = \sigma \phi w N_A \theta (1-\exp(-\lambda t))/A$$

Usually in neutron activation analysis, the activity of the radionuclide is measured experimentally in a sample to deduce the unknown mass of the element using the activation equation:

$$w = A_0 A/N_A \sigma \phi \theta (1 - \exp(-\lambda t))$$

Corrections must also be made for the decay period t_d before counting:

$$w = A_0 A/N_A \sigma \phi \theta (1 - \exp(-\lambda t)) \exp(-\lambda t_d)$$

NUCLEAR INTERFERENCES

We have seen in the previous section that stable isotopes of an element can undergo neutron activation to produce several different nuclei depending on the energy of the neutron flux. It is therefore possible for competing reactions to take place via different nuclear reactions to give the same product nucleus. For example, in a rock sample containing aluminum, silicon and phosphorus the following (n,γ), (n,p) and (n,α) reactions may occur in a neutron flux consisting of both thermal and fast neutrons:

$$^{27}_{13}Al + n \rightarrow ^{28}_{13}Al + \gamma$$

$$^{28}_{14}Si + n \rightarrow ^{28}_{13}Al + p$$

$$^{31}_{15}P + n \rightarrow ^{28}_{13}Al + \alpha$$

The ^{28}Al activity in this case may be due to aluminum, silicon and phosphorus in the sample. The effect of the interference will depend on the cross section of each of the target nuclei and on the size of the thermal and fast components of the neutron flux. Other examples of competing reactions include:

$$^{51}V(n,\gamma)^{52}V \text{ and } ^{52}Cr(n,p)^{52}V$$

$$^{41}K(n,\gamma)^{42}K \text{ and } ^{42}Ca(n,p)^{42}K$$

$$^{59}Co(n,\gamma)^{60}Co \text{ and } ^{63}Cu(n,\alpha)^{60}Co$$

When choosing a suitable activation product for analysis it is important to ensure that there are no competing reactions which may affect the result.

Another nuclear reaction which is a common source of interference is fission. When, for example, uranium undergoes fission a large number of fission products may be produced. Some fission products are very unstable and decay rapidly to other radionuclides. The predominant fission products tend to be those with masses around 95 and 140, such as ^{95}Zr, ^{103}Ru, ^{137}Cs, ^{140}Ba and ^{144}Ce. Radionuclides produced by (n,γ) reactions with the rare earth elements in particular are affected by interference from fission products. For example ^{140}La, ^{141}Ce, ^{147}Nd, and ^{153}Sm are all produced from (n,γ) reactions and via uranium fission. The problem is particularly significant in the analysis of uraniferous rocks for the lanthanides (Ila et al., 1983).

Neutron self-shielding can also cause interferences in the activation of a target nucleus. This occurs when an element in the sample matrix absorbs neutrons in competition with the target of interest. For example, a material with a high cross section to thermal neutrons, such as boron, would absorb and therefore reduce the thermal neutrons from the source such that the target would be seeing a lower neutron flux and be activated to a lesser extent than expected. This effect is greater the larger the mass of the sample and can be detected by irradiating a range of sample masses and calculating the induced activity per unit mass. If neutron self-shielding is occurring, then as the mass increases so the activity per unit mass decreases.

REFERENCES

Browne, E. and R. B. Firestone (1986), *Table of Radioactive Isotopes*, Wiley, New York.

Bujdoso, E., I. Feher, and G. Kardos (1973), *Activation and Decay Tables of Radioisotopes*, Elsevier, Amsterdam.

Ila, P., P. Jagam, and G. K. Muecke (1983), "Multielement analysis of uraniferous rocks by INAA: special reference to interferences due to uranium and fission of uranium," *J. Radioanal. Chem.*, **79**(2), 215–232.

Mughabghab, S. F., M. Divadeenam, and N. E. Holden (1984), *Neutron Cross Sections, Volume 1: Neutron Resonance Parameters and Thermal Cross Sections. Part A: z=1–60*, Academic Press, New York, and *Part B: z=61–100*, Academic Press, Orlando, Florida.

CHAPTER

3

IRRADIATION FACILITIES

The choice of neutron source for activation analysis will be dictated by the energy range of neutrons required and in some cases, because of the specialized nature of the equipment, what is available and accessible. In general the majority of useful activation products are the result of interactions with thermal neutrons and the highest neutron flux is obtained using a thermal nuclear reactor. The reactor will have a high component of thermal neutrons, in the order of 10^{14}–10^{18} n m^{-2} s^{-1}, but there will also be a significant proportion of epithermal and fast neutrons. Epithermal neutrons may be used to enhance the activation for some elements and irradiations are made with a thermal neutron filter of cadmium or boron in the irradiation site. However there are some useful reactions with fast neutrons and the second most common choice of source is a neutron generator, which produces monoenergetic neutrons of 14 MeV. The generator has an advantage over the reactor in that it is portable and easily operated. Finally there are other smaller neutron sources such as ^{252}Cf or Am/Be which are useful for specific applications. The irradiation sites in a neutron source have a wide range of geometries which dictate the shape and size of the sample container. The variation in neutron flux within the irradiation site is also a significant factor and must be evaluated and corrected for if necessary.

NUCLEAR REACTORS

There are many designs of reactors but the principles on which they are based are very similar. Fissionable material will undergo fission with neutrons, accompanied by the production of further neutrons. Most of the reactors used for neutron activation analysis are based on the fission of ^{235}U, where the reactor core is made up of fuel elements of uranium enriched in ^{235}U. ^{235}U will only undergo fission with thermal neutrons and so the core is surrounded with a moderator to slow the neutrons down. The moderator is usually in the form of light water, heavy water, graphite or beryllium. There is a lot of heat generated during the process and so there is also a coolant which may be a gas, water or liquid metal. In a "swimming pool" reactor, for example, light water may be used both as the moderator and as the coolant. Control rods, made of a neutron-absorbing material such as cadmium or boron, are used to control the rate of production of neutrons. The reactor is allowed to reach its operating power by gradual removal of control rods from the core. The reactor is maintained at a constant operating power by adjustment of a fine control rod and the reactor is shut down by moving the control rods back into the core.

Typical reactors used for neutron activation include the Canadian Slowpoke® reactor, the American TRIGA® reactor and other pool type reactors. The construction of the Slowpoke® reactor is shown in Figure 3.1. It operates at 20 kW and is designed specifically as a teaching tool with activation analysis and production of small amounts of radioisotopes in mind. The fuel is 19.9% enriched uranium, consisting of 200 fuel elements in a compact core measuring 220 mm diameter by 227 mm high. The system is designed to operate remotely. The reactor can be provided with up to five irradiation sites in the maximum flux of 10^{16} n m^{-2} s^{-1} and five further tubes outside the reflector with a maximum flux of 0.5×10^{16} n m^{-2} s^{-1}. The TRIGA® reactor is designed as a general purpose reactor but has incorporated in its design the facility for flux equalization systems which make it very suitable for activation analysis.

The majority of the neutrons in a pool type reactor are thermalized and at these energies the activation process, with interaction of a neutron and absorption in the nucleus, is most likely to occur. Most of the trace elements of interest may be detected by activation of a stable nucleus with thermal neutrons. Although most of the neutrons in the reactor core are slowed down to energies below the cadmium cutoff energy of 0.5 eV, there is still a component with higher energies, in the epithermal and fast region. Activation at neutron energies above the cadmium cutoff occur as resonances, where only neutrons with a specific energy will be absorbed to activate the target nuclei. It is possible to slow down the neutrons from the reactor by placing the target in a site surrounded by a moderator such as graphite. Similarly thermal neutrons can be removed from the neutron flux using a strong thermal neutron absorber such as cadmium or boron, to provide a region of epithermal and fast neutrons.

Many reactors are unique in their design. The maximum power outputs of nuclear reactors range from kW to MW. Correspondingly the neutron flux obtainable ranges up to a maximum of about 10^{18} n m^{-2} s^{-1}. However, the neutron flux in a particular irradiation site will depend on its distance from the reactor core. The higher the neutron flux in neutron activation, the better the activation of the sample but the activation of

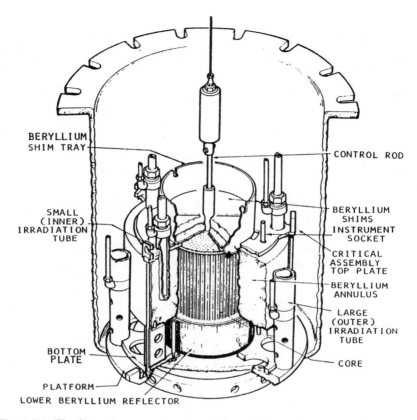

BERYLLIUM
SHIM TRAY

CONTROL ROD

SMALL
(INNER)
IRRADIATION
TUBE

BERYLLIUM
SHIMS
INSTRUMENT
SOCKET

CRITICAL
ASSEMBLY
TOP PLATE

BERYLLIUM
ANNULUS

LARGE
(OUTER)
IRRADIATION
TUBE

BOTTOM
PLATE

CORE

PLATFORM

LOWER BERYLLIUM REFLECTOR

Figure 3.1. The Slowpoke reactor: a view of the critical assembly and reactor container lower section. (Reprinted with the permission of AECL Research, Canada.)

the background interferences increases similarly so this means that the improvement in detection of the element of interest is not direct. The main consideration when choosing the reactor to use for activation is that there are suitable sites for irradiating samples and that they are accessible. Realistically, there is often very little choice of irradiation site available and the user may have to accept the limitations of the facility.

IRRADIATION CONTAINERS

The samples to be irradiated for activation analysis must be contained in some way. The material of the container should not itself become too radioactive otherwise it will create a hazard to handlers. Polyethylene is generally used for thermal neutron irradiations since high purity grade polyethylene contains very few impurities that activate significantly with thermal neutrons. Such polyethylene capsules are readily available from laboratory suppliers but they do not always have the correct dimensions for a particular job. In that case it is quite simple to get low density polyethylene capsules injection-molded with a purpose-made die. Polyethylene can also be easily heat sealed using a conventional soldering iron which makes preparation simple. If necessary it is also possible to machine polyethylene to produce lids which screw on, although it will add to the cost. Because polyethylene is relatively pure it is often possible to count the activated capsule with its contents, without transferring the sample to a new container. This could be a great advantage if it avoids handling radioactive powders, for example.

The main disadvantage of polyethylene is the radiation damage that occurs on irradiation in a high integrated neutron flux. The polymer deteriorates to a brittle material which is discolored and in extreme cases it will crack and disintegrate. It is common to use aluminum cans with screw caps for prolonged irradiations in place of polyethylene. Aluminum activates to give ^{28}Al, with a half-life of 2.3 min, which may be a radiation hazard on short irradiations but not after a relatively short decay period. Irradiation of liquids can present a problem and the way in which the irradiation capsules are sealed will be particularly important. However, there is no reason why liquids cannot be irradiated, provided that attention is paid to the possible evolution of gases due to radiolysis. This effect may limit the amount of liquid or the length of irradiation. The temperature in the irradiation site may have a greater impact on a liquid sample and that again will depend on the reactor. Quartz glass ampoules are normally used for irradiating liquids when they are subsequently to undergo radiochemistry to facilitate the complete transfer of the sample. The purity of the glass is then very important as it will be analyzed along with the sample.

Whatever the nature of the particular irradiation device the irradiation container will probably come into contact with radioactive dust from the irradiation tube. Therefore it is advisable to pack the sample in an inner capsule, so that the sample container is not contaminated with the dust. The reactor operations staff will probably insist on double-containment from a safety point of view in case of leakage from the inner container. If more than one sample capsule may be packed into the irradiation container it will be possible to combine a number of samples and/or standards together in one irradiation container. In addition, where a stacking device is used, it may be possible to load a number of containers into the same outer irradiation tube.

The shape and size of the irradiation containers will depend on the irradiation site that they have to fit. The irradiation tubes are usually cylindrical and the inner diameter of the irradiation tube will dictate the outer diameter of the irradiation container. If a pneumatic transfer system is used, where the container is blown under pressure along a tube, then the shape becomes critical as well as the size. Pneumatic systems are more sensitive to capsule shape because the diameter of the container (in this case often referred to as the rabbit) must be close to the inner bore of the pneumatic tube so that the gas does not escape round the gap. However, the capsule must be able to travel round any bends in the system. The shapes of the rabbits designed to overcome these problems can be quite different, some have a reduced profile at the ends whereas others have a ring at the ends to enhance the flight. Figure 3.2 shows examples of the shapes of containers used for different irradiation sites.

IRRADIATION DEVICES

Introducing a sample into a neutron source requires care and although this section refers to the problems associated with reactor irradiations, the same difficulties will apply to other neutron sources. It is often the case that neutron activation analysis laboratories have been introduced after the reactor itself has been installed and operating for a while. The irradiation facilities are developed or modified to cater for activation analysis and as a result they vary even between reactors with the same original design. The way of introducing a target into the neutron flux will depend on the physical structure of the reactor. It is imperative that the introduction of a sample into the reactor vessel does not affect the operation of the reactor. The aim is usually to get the sample close to the reactor core, without affecting its operation.

The irradiation site may be within the core itself or close to the outside edge of the core configuration. In a pool type reactor the irradiation tubes will be in water, if the moderator is graphite then the irradiation tubes will be located in the graphite block. If the reactor can be accessed from above then vertical tubes may be installed to lower the samples down so that they are in a final position close to the side of the core or inside it. Alternatively, horizontal tubes can be installed, terminating close to the core. Closed tubes are used to house the samples so that they are not actually in contact with the water surrounding the core. The tubes are usually made of aluminum because of its resistance to corrosion

polyethylene 'rabbits' for pneumatic systems

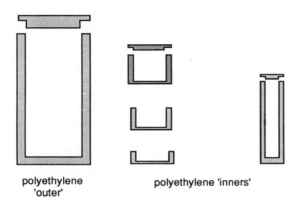

polyethylene   polyethylene 'inners'
'outer'

cylindrical capsules for stringer devices

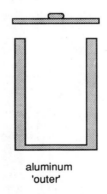

aluminum
'outer'

Figure 3.2. Designs of polyethylene containers used for reactor irradiations.

and the fact that activation products are short-lived, so if they do have to be removed from the reactor there is little residual radiation. The aluminum itself has a small effect on the neutron flux.

Since the aim is to position the sample close to the core, the final location must be done remotely. Samples can be loaded into the irradiation site manually by lowering them down vertical tubes until they are in line with the reactor core. Figure 3.3 shows a typical stringer device. Ladder type racks are used to stack the samples one above the other, along the length of the core, and then they are lowered using, for example, nylon string, to the bottom of the irradiation tube. Alternatively, if a horizontal

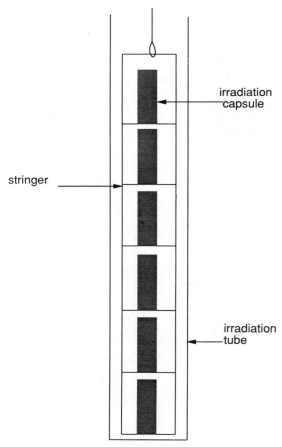

irradiation capsule

stringer

irradiation tube

Figure 3.3. A simple stringer device used to lower samples into the reactor irradiation tube.

irradiation tube is used, the samples may be pushed into position using a locating rod. If the irradiation time is short, say a few minutes or less, it is important to have a reproducible time for loading and unloading samples. The manual loading of samples is neither quick nor reproducible and a mechanical system is necessary. Mechanical loading devices include chain-driven racks which carry the samples to their final destination, a relatively slow way to travel but reproducible. A sample loaded with a device which is slow will travel through a region of low to high neutron flux to reach its destination and when it is unloaded will go back through the high to low flux. Consequently it is seeing an integrated neutron flux which is difficult to quantify but which will be reproducible each time a sample is irradiated.

If it is important that the sample is loaded and unloaded rapidly, as in the case of very short irradiations, it is more convenient to use a pneumatic device. In a reactor environment it is valuable to keep moving parts to a minimum and the use of gas to move the sample has several advantages. Figure 3.4 shows a pneumatic device of typical design. The sample is propelled along the irradiation tube with pressurized air or nitrogen gas and it is quite straightforward to irradiate samples this way using transit times of a second or less. A pneumatic irradiation system

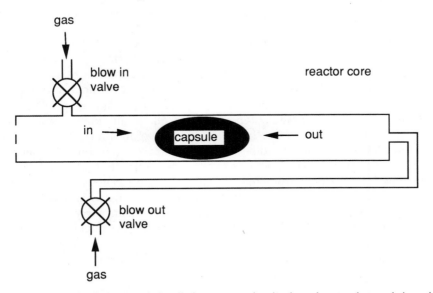

Figure 3.4. A simple pneumatic irradiation system, showing how the capsule travels in and out of the reactor core depending on direction of the gas flow.

consists of three stages: the sample is loaded into the tube and blown into the irradiation site, it is then left for the required irradiation time and then it is blown out again. The sample is blown in by applying compressed gas to the end of the irradiation tube. To remove the sample the gas is applied to the end of the irradiation tube via a separate inlet or by the use of concentric tubing, blowing the gas down the outer tube to blow the sample out. The flow of the compressed gas is controlled using a solenoid valve operated by applying the power for the time required for transit, usually a second or two. The operation of the solenoid valves can be controlled manually.

A pneumatic device can be automated very readily and the process may be implemented in stages. The first stage in automation would be to control the length of irradiation with a clock which is initiated once the blow-in valve has been operated to send the sample into the irradiation site. Once the preset irradiation time is reached the blow-out valve is triggered to send the sample back out of the reactor. This allows a single sample to be irradiated in a very reproducible way. The next stage is to irradiate a number of samples in sequence by introducing a sample loader. The loader could be used to stack a number of samples and introduce them one by one into the tube to be blown into the irradiation site. A counter could monitor the number of samples that had been irradiated, stopping at the preset number, or the loader could simply signal when it is empty. Finally a waiting period can be introduced between irradiations, if a gap is required.

Automatic control of the solenoid valves and time intervals can be based on the use of microprocessors. Programmable logic controllers, which have been developed for process control applications to replace logic circuitry, are suitable controllers. They consist of input and output modules, a central processor and a programming terminal, providing a cheap and very flexible system for controlling automatic irradiation systems. If the sample is to be analyzed very soon after irradiation, as in the case of short-lived radionuclides and delayed neutrons, the sample can be transferred automatically to the counting station, again using pneumatics. A signal from the automatic system triggers the counting equipment when the sample is in the counting position. An example of a typical automated system, used for delayed neutron analysis, is shown in Figure 3.5. Figure 3.6 is a schematic diagram of a pneumatic system for gamma ray analysis, where there is an option to send the sample back into the reactor after counting for reirradiation, the so-called cyclic activation system (Burholt et al., 1982).

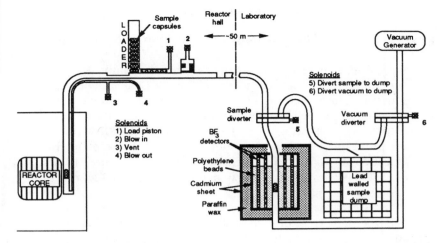

Figure 3.5. An automated pneumatic irradiation system for delayed neutron analysis. (Reprinted with permission from Benzing and Baghini, 1990.)

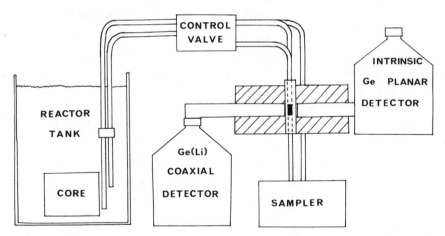

Figure 3.6. An automated pneumatic irradiation system for gamma ray analysis where the same sample can travel from the reactor to the detector many times for cyclic activation analysis.

29

EPITHERMAL NEUTRON ACTIVATION

In a thermal nuclear reactor it is expected that the major portion of the neutrons will be thermal but there will be a smaller component of epithermal and fast neutrons. This does not pose a problem unless a neutron flux free of higher energy neutrons is required. A thermalized neutron irradiation site can be created in a reactor away from the reactor core and surrounded with a good moderator such as graphite. Because of its location away from the high flux close to the core the thermal column will have a lower thermal neutron flux than the rest of the irradiation sites.

In some cases there may be a requirement for epithermal neutron activation, perhaps to reduce an interfering reaction induced by thermal neutrons. The thermal neutrons are filtered out using a thermal neutron absorber, usually made from cadmium or boron; both are very good thermal neutron absorbers. Cadmium 1 mm thick is "black" to thermal neutrons and the sample inside a 1 mm thick cadmium filter sees only the neutrons above 0.5 eV and boron cuts out neutrons with a higher energy, at a point which is dependent on the thickness of boron or boron compound.

It is possible to install cadmium-lined irradiation tubes in a reactor to facilitate these epithermal neutron irradiations. The epithermal irradiation tube is made by wrapping the 1 mm cadmium sheet around the aluminum irradiation tube and then using an outer aluminum layer to cover the tube so that the cadmium is not in contact with the moderator. Boron and its compounds are more difficult to machine and so it is not so suitable for installation in the reactor vessel. Several cadmium-lined tubes have been added to the Consort reactor in the UK and Figure 3.7 shows the layout of the irradiation sites, including three different types of pneumatic system and three different sizes of cadmium tube.

In high power reactors there may be a problem of burn up in a cadmium tube and it may not be practical to install an epithermal activation tube in the reactor. Besides, if space is limited it may be a luxury to have an irradiation site devoted to epithermal neutron activation. It is possible to construct an irradiation capsule with cadmium or boron walls for loading samples into. Cadmium-lined irradiation cans are also used and even combinations of both cadmium and boron. When irradiated, cadmium produces significant levels of activity which can cause handling problems if the cans are opened soon after irradiation and boron compounds are preferred in these cases. Since the activation products give negligible radioactivity, boron carbide or boron nitride are used to fabricate irradiation containers. The design of can used for epithermal

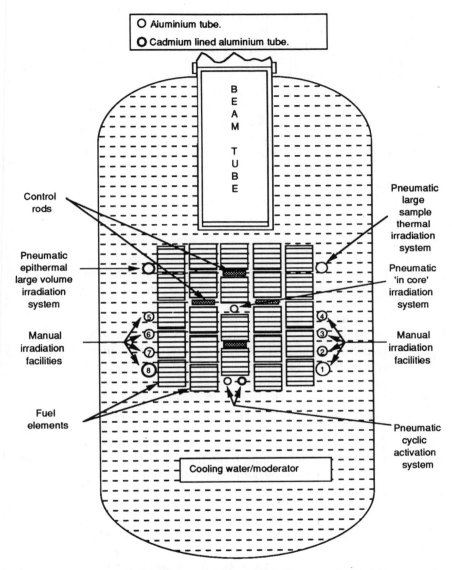

Figure 3.7. A schematic diagram of the Consort Mark II reactor showing the pneumatic and manual irradiation facilities for both thermal and epithermal neutron activation analysis.

31

neutron activation is illustrated in Figure 3.8 (Cesana et al., 1978; Glascock et al., 1985; Parry, 1982).

Clearly for the measurement of very short-lived radionuclides which must be counted immediately it is preferable to avoid the irradiation can altogether and use an epithermal irradiation site with the absorber in the tube installed in the reactor. An additional problem with absorber cans is that there is usually a limit to the surface area of such neutron absorbers that may be loaded into the reactor because of the effect that it has on the reactivity. This effect is due to depression of the neutron flux caused by the presence of the absorber. When the absorber is removed from the reactor, the neutron flux will rise. This will cause a particular problem where the sample is loaded and unloaded at power and consequently the amount of absorber will be limited to a safe level.

NEUTRON FLUX DISTRIBUTION

The distribution of the neutron flux within a neutron source is not necessarily constant. On the contrary, in a reactor the neutron flux is unlikely to remain constant even over relatively short distances. The neutron flux falls off with distance in both vertical and horizontal directions away from the center of the reactor core. The effect is a cosine curve with the maximum at the core centerline. Neutron activation analysis is based on the principle that induced activity is proportional to neutron flux and therefore any quantitative measurements rely on the neutron flux being reproducible or quantifiable.

The best way to do this is to keep the flux as constant and reproducible as possible. On the TRIGA® reactor this is achieved using a rotating irradiation device, where the samples travel slowly round the perimeter of the core so that all the samples see the same integrated flux during an irradiation. The main disadvantage of equalization systems is that by their nature the samples will all see an average flux, obtained by traveling through regions of high and low neutron density. The compact nature of some reactor vessels means that there is little space available for equalization systems if maximum neutron fluxes are to be achieved.

If it is not possible to use a flux equalization system, it is necessary to accept the variations that exist, carefully measure them and make the necessary correction for them. The neutron flux distribution in an irradiation site can be characterized using a neutron flux monitor, either at the time of the experiment or in a separate measurement. Normally the neutron flux in an irradiation site is monitored using an activated foil. Care must be taken in the selection of the material used since an

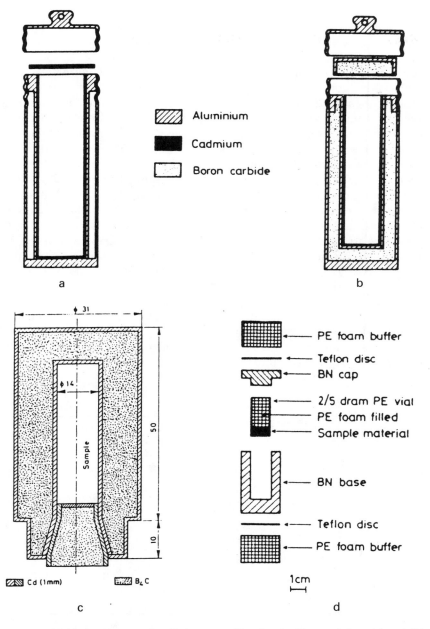

Figure 3.8. Epithermal neutron irradiation cans. (Reprinted with permission, (a) and (b) from Parry, 1982; (c) from Glascock et al., 1985; and (d) from Cesana et al., 1978.)

element activated only by thermal neutrons will only give an accurate measure of the thermal neutron flux in the irradiation site. In some cases the element to be measured may be activated by epithermal neutrons and so the epithermal neutron component must also be measured by another means. So not just one, but two monitors are needed: a mainly thermal neutron absorber such as cobalt, copper or iron and a mainly epithermal neutron absorber such as gold. A combination monitor of cobalt and gold, for example, will provide all the necessary information.

The distribution of the neutron flux along a core tube is shown by Figure 3.9. Cobalt was used as the flux monitor. The activation follows the cosine curve expected for an irradiation site close to and along the length of the reactor core. The positions where the irradiation capsules are placed are indicated on the graph and clearly the variation is much less close to the centerline of the core. At the extreme ends of the site the variation in neutron flux even along an irradiation container can be as high as 20%. It is therefore extremely important that the variation is monitored and the proper correction made for the difference in the neutron flux experienced by adjacent samples.

The problem of flux variation due to distance will be minimized if all samples are activated in the same irradiation site and in the same position within an irradiation capsule, without stacking samples. This is only possible if samples are irradiated singly, which is often the case for short

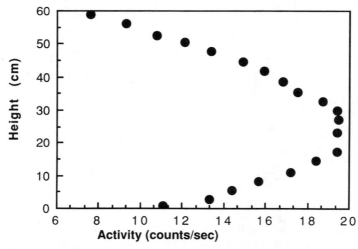

Figure 3.9. Neutron flux variation along an irradiation tube, demonstrated by the activity induced in a cobalt monitor.

irradiations when a pneumatic irradiation device is used and for other neutron sources such as accelerators and neutron generators. In this way the samples are irradiated in sequence, one after another and so the variation of the neutron flux with time becomes important. The operators of research reactors usually maintain a constant power. However if the use of the reactor involves changes in core configuration there could be dramatic changes in the neutron flux in an irradiation site from one day to the next. Even a small change such as the introduction of a cadmium-lined irradiation can or some other strong neutron absorber will depress the thermal neutron flux in close proximity to it.

OTHER NEUTRON SOURCES

Neutron sources not involving the nuclear reactor, namely accelerators and isotopic sources, are used in only a small proportion of applications in neutron activation analysis. However, these sources are often used in commercial applications and their contribution to neutron activation analysis may be greater in terms of the number of samples analyzed. Often seen as complementary to nuclear reactors, which have predominantly thermal neutron fluxes, the accelerator group is used to produce more energetic neutrons. Isotopic sources are very useful because they do not involve high voltage units and are small, compact and mobile.

Commercially available D–T neutron generators, based on the reaction of a deuteron beam on tritium $^3H(d,n)^4H$, are generally used for fast neutron activation analysis. The hydrogen ions are accelerated to 150 kV and deuterium ion currents are in the range of 1–5 mA. The resulting neutron energy is approximately 14 MeV and typically the neutron source strengths are of the order of 10^{11} n s^{-1}. The fast neutron generator is of specific use in the determination of carbon, sulphur and silicon and has the advantage that it can be installed on an industrial site. For example a fully automated system is used by an oil company utilizing a fast neutron generator for the analysis of oxygen in oil products. It processes about 1600 samples per year (Filpus-Luyckx and Ogugbuaja, 1987).

The Van de Graaff generator and cyclotron are based on the action of a deuteron beam on a target of beryllium, for example: $^9Be(d,n)^{10}Be$. There has not been widespread application of the technique for neutron activation analysis but the use of a cyclotron is examined and detection limits given for many elements in recent work by Esprit et al. (1985) where the deuteron beam was accelerated up to 14.5 MeV and the beam intensity was 20 μA. Electron accelerators such as the betatron, synchrotron, the microtron or a linear accelerator, produce high energy bremsstrahlung, resulting in the (γ,n) reaction in the target. An example of the routine use of a linear accelerator is given by Ivanov et al. (1985) who describe the application of an 8 MeV electron accelerator to the determination of gold at a mining installation in the Soviet Union, where it is used to process some 250,000 samples per year.

The use of isotopic sources in neutron activation analysis was recently reviewed in an article by Garg and Batra (1986), which includes many references to the use of these sources for the analysis of ores, alloys and industrial material. There are basically three types of isotopic source used in activation analysis: photoneutron (γ,n) sources, alpha (α,n) sources and spontaneous fission sources.

The photoneutron source is based on the fact that ^{9}Be in particular releases neutrons when bombarded with gamma rays. ^{2}D is also useful as a photoneutron source. ^{9}Be is chosen because the threshold for the reaction is at 1.67 MeV, whereas for ^{2}D it is at 2.33 MeV and all other nuclides have a threshold above 6 MeV. Consequently using ^{9}Be the reaction may occur with a gamma ray source such as ^{124}Sb, which emits a gamma ray at 1.7 MeV. The source itself is produced by surrounding the gamma emitter with a thick layer of beryllium. The energies of the neutrons emitted by these photonuclear sources are of the order of 0.2–0.8 MeV, depending on the target, and of course they may be moderated with a layer of paraffin wax or polyethylene.

Alpha sources consist of Be mixed in Po so that it is activated with alpha particles in the reaction:

$$^9\text{Be} + {}^4\text{He} \rightarrow {}^{12}\text{C} + {}^1\text{n} + 5.7 \text{ MeV}$$

Other targets include B, F and ^{13}C, but Be gives the greatest number of neutrons per alpha particle and the energy can be high, ranging from 6.7–11 MeV depending on the angle of scatter.

The isotopic source based on the spontaneous fission of ^{252}Cf has found wide applicability. It has a half-life of 2.6 y so it has a useful working life. The average neutron energy is 2.348 MeV, which may be moderated. An example of its commercial use is a system used for the routine analysis of crude oil samples for vanadium. The ^{252}Cf source consists of a 6 GBq source (2.9 mg), shrouded in stainless steel in a large water tank, with a maximum thermal flux estimated at 10^{12} n m^{-2} s^{-1}. It is capable of processing some 10,000 samples per year (Lubkowitz et al., 1980). The californium source is valuable for use in the field, for example it is used as an on-line analyzer for iron and aluminum in iron ore fines. The diagram in Figure 3.10 shows how the sample is analyzed

37

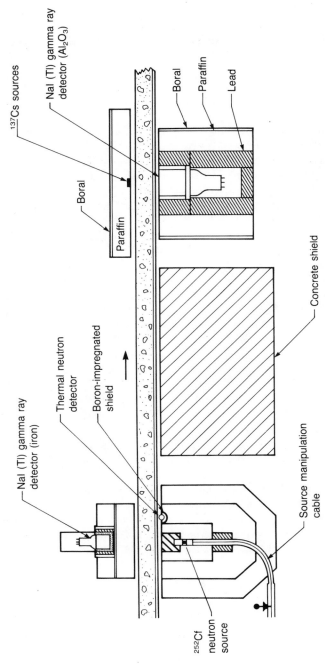

Figure 3.10. On-stream analyzer for iron ore fines. Iron is measured at the point of irradiation and aluminum is monitored 2 m down the line. (Reprinted from Holmes et al., 1980, courtesy of R. J. Holmes, CSIRO, Australia.)

during irradiation for iron and then 2 m down the line for aluminum (Holmes et al., 1980).

REFERENCES

Benzing, R. and N. M. Baghini (1990), "Imperial College Reactor Centre 1990," *Imperial College Reactor Centre Report*, Ascot, UK.

Burholt, G. D., E. A. Y. Caesar, and T. C. Jones (1982), "The fast cyclic activation system for neutron activation analysis in the University of London Reactor," *Nucl. Instr. Meth.*, **204**, 231–234.

Cesana, A., F. Rossitto, and M. Terrani (1978), "On the resonance activation analysis of biological materials," *J. Radioanal. Chem.*, **45**, 199–202.

Esprit, M., C. Vandecasteele, and J. Hoste (1985) "Sensitivities and interferences in activation analysis with cyclotron-produced fast neutrons," *J. Radioanal. Nucl. Chem.*, **88**, 31–44.

Filpus-Luyckx, P. E. and V. O. Ogugbuaja (1987), "An automated pneumatic transfer system for oxygen determinations by neutron activation analysis," *Nucl. Instr. Meth. Phys. Rev.*, **B24/25**, 1017–1020.

Garg, A. N. and R. J. Batra (1986), "Isotopic sources in neutron activation analysis," *J. Radioanal. Nucl. Chem.*, **98**, 167–194.

Glascock, M. D., W. Z. Tian, and W. D. Ehmann (1985), "Utilization of a boron irradiation vessel for NAA of short-lived radionuclides in biological and geological materials," *J. Radioanal. Nucl. Chem.*, **92**(2), 379–390.

Holmes, R. J., A. J. Messenger, and J. G. Miles (1980), "Dynamic trial of an on-stream analyser for iron ore fines," *Proc. Australas. Inst. Min. Metall.*, **274**, 17–22.

Ivanov, I. N., O. K. Nikolaenko, and A. S. Shtan (1985), "Methods and equipment of industrial activation analysis," *J. Radioanal. Nucl. Chem.*, **90**(1), 189–196.

Lubkowitz, J. A., H. D. Buenafama, and V. A. Ferrari (1980), "Computer controlled system for the automatic neutron activation analysis of vanadium in petroleum with a californium-252 source," *Anal. Chem.*, **52**, 233–239.

Parry, S. J. (1982), "Epithermal neutron activation analysis of short-lived nuclides in geological material," *J. Radioanal. Chem.*, **72**, 195–207.

CHAPTER

4

GAMMA RAYS

The principle of activation analysis is to induce radioactivity in the element of interest. Decay products are then detected and identified by their energies while the decay rate is used to make a quantitative determination of the target element. Consequently it is the aim of the analyst to choose the most suitable conditions for identification of the product and sensitivity of measurement. In general gamma rays are the most suitable form of radiation for multielemental analysis. The best gamma ray lines are selected on their abundance which is defined from the mode of decay. The decay schemes are used to find the branching ratios or gammas per disintegration. Finally the gamma rays are detected with a semiconductor detector, based on the interaction of the photon within a pure material.

RADIOACTIVE DECAY

The types of radioactive decay products which are likely to occur and which may be used to identify and measure radioactivity are alpha and beta particles, gamma and X-rays and neutron emission.

Alpha particles are emitted during the course of the decay chain of naturally occurring radioactive species and are not usually emitted as the result of decay of any neutron activation process, except where the activation process results in the production of a nuclide in the natural decay chain, for example ^{235}U decaying to ^{231}Th with emission of an alpha particle or, in the case of neutron-induced reactions of lithium and boron, $^{6}Li(n,\alpha)^{3}H$ and $^{10}B(n,\alpha)^{7}Li$, when prompt alpha particles are emitted. Alpha particles are monoenergetic and have energies in the range 2–10 MeV. The range of an alpha particle is small, for example a 5 MeV alpha particle will travel only 0.03 mm in silicon. Consequently, before counting, an alpha emitter would normally be chemically isolated from the original sample matrix and plated onto a support so it could be measured. For these reasons alpha particles are rarely used in activation analysis for the determination of elemental concentration. Notable exceptions are the determination of uranium by measuring the alpha particles produced by fission during neutron activation, and the determination of boron in steel. The alpha particles are recorded using a film made of cellulose nitrate, cellulose acetate or a polycarbonate placed in close contact with the surface of the material under examination. When an alpha particle is emitted it creates tracks in the film. The number of tracks recorded will indicate the concentration of the alpha emitter in the sample.

Beta particles are charged particles with either a negative (β^-) or a positive (β^+) charge. They are not monoenergetic but have a band of energy from zero up to a maximum energy in the range 0.1–20 MeV, although the majority are below 10 MeV. Most neutron-activated nuclei decay by emission of beta activity giving rise to a daughter with either an atomic number one above, in the case of β^- emission, or one below, in the case of β^+ emission. For example, the decay of ^{60}Co with emission of a β^- produces ^{60}Ni and the decay of ^{65}Zn with emission of a β^+ gives ^{65}Cu. The major disadvantage is the wide band of energies exhibited by beta particles. A mixed source containing several beta emitters would give a range of energies which could only be identified by their different half-lives and by chemical separation of the components. This work is applied to some analytical problems and can provide a very sensitive method of analysis but since it is a specialized technique usually confined to single elements, it is not suitable for multielemental analysis.

X-rays are produced as a result of internal conversion or electron capture during the decay of radionuclides produced by neutron activation. In the case of electron capture the nucleus captures an electron from the K or L shell, with the production of a daughter with an atomic number one less and emission of the characteristic X-ray of the daughter. For example, ^{58}Co decays by electron capture and positron emission to give ^{58}Fe and the characteristic X-ray of iron. Internal conversion occurs when a gamma ray emitted from the nucleus interacts with an orbital electron producing a conversion electron. Rearrangement of the orbital electrons results in production of an X-ray characteristic of the parent. The percentage of internal conversion is often high for isomeric transitions, such as the decay of ^{60m}Co to ^{60}Co, when virtually all the gamma rays (59 keV) are converted in the K shell and the characteristic X-ray of cobalt is seen. The X-rays produced during these decay processes may be used in activation analysis to measure elements. Being monoenergetic they are suitable for multielemental analysis. The main problems of X-ray counting are associated with interferences from other processes, secondary X-rays due to fluorescence, high backgrounds due to beta particles and absorption in the matrix. The latter is a major drawback for analyzing thick samples since the attenuation effects of X-rays, with energies between a few and 100 keV is large. It can be overcome using very thin samples. Activation analysis with X-rays and their applications is described in a review by Mantel and Amiel (1981).

Gamma rays are produced as a result of a transition between excited levels of a nucleus. These may be prompt gamma rays produced when a neutron is absorbed during activation of the (n,γ) type, or a delayed gamma ray which occurs when the radioactive product is decaying. In the first case the gamma rays are emitted during irradiation, as the highly excited level decays to the zero level of the activation product. This transition can only be seen while the sample is being irradiated. Prompt gamma ray analysis is a method of activation analysis which has specific applications and has been reviewed by Peisach (1981). The gamma ray spectra tend to be quite complex and have a number of very high energy gamma rays (up to 6 MeV). Delayed gamma rays are emitted during the decay of the parent and normally follow the beta decay. Therefore they have the characteristic half-life of the parent. The transitions are between energy levels of the daughter and therefore there may be a number of different gamma ray lines. Each photon has a characteristic well-defined energy and there may be several energy transitions for each disintegration. Typical energies are in the range 0.05–4 MeV.

When a fissionable material (uranium, thorium, neptunium or plutonium) is activated with neutrons to induce fission, short-lived fission products with half-lives of up to 54 s are produced. They are very unstable and decay sometimes with emission of a neutron from the nucleus. These delayed neutrons have an average energy of 0.5 eV. They provide a specific technique for measuring fissionable material. The method is applied in particular to the determination of uranium in mineral exploration samples (Amiel, 1981). It is a very specific technique and its applications are limited to fissionable material.

DECAY SCHEMES

The choice of radionuclide for use in activation analysis will depend on the nature of its decay. The way in which radioactive species decay is illustrated by the use of decay schemes which are found in the *Table of Isotopes* (Lederer and Shirley, 1978). For example, the scheme shown in Figure 4.1 represents the decay of ^{35}S, which may be produced from the neutron activation of ^{34}S. It is a pure beta emitter and decays with a half-life of 88 days to the stable daughter. The daughter produced as a result of beta emission has one proton more than the precursor, which

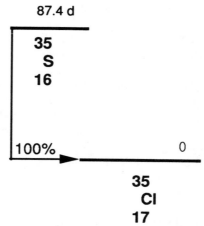

Figure 4.1. Decay scheme for 35S. (Reproduced with permission from Lederer and Shirley, 1978.)

is ³⁵Cl. In the absence of any gamma ray lines in this case it would only be possible to measure the sulphur content with the beta activity (0.167 MeV).

An example of a decay scheme involving a gamma ray transition is represented in Figure 4.2, for ²⁸Al produced as a result of the ²⁷Al(n,γ) reaction. In this case the decay of the ²⁸Al with a half-life of 2.24 min is via emission of a beta particle, with a maximum energy of 2.88 MeV,

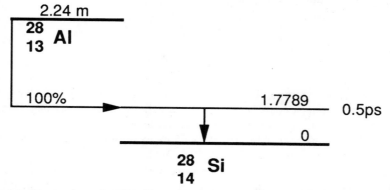

Figure 4.2. Decay scheme for ²⁸Al. (Reproduced with permission from Lederer and Shirley, 1978.)

to the 1.78 MeV energy level of the daughter ^{28}Si. The gamma transition to the stable state of the daughter results in the emission of a gamma ray with an energy of 1.78 MeV. The lifetime of the transition state is only 0.5 ps, so the decay is effectively at the rate of the half-life of the ^{28}Al.

Few of the decay schemes are so simple and there are often a number of gamma transitions for a particular decay. In the case of ^{46}Sc, shown in Figure 4.3, there are two gamma ray lines. Neutron activation of scandium (100% ^{45}Sc) results in the production of ^{46m}Sc and ^{46}Sc. The short-lived ^{46m}Sc is the energetic level which decays by isomeric transition with emission of a gamma ray with an energy of 143 keV. This will be accompanied by internal conversion and the characteristic X-ray of scandium will also be produced. ^{46}Sc decays by beta emission almost completely to the 2.01 MeV energy level. The subsequent transition to the 0.889 MeV level results in the emission of a 1.12 MeV gamma ray, which is followed by a 0.889 MeV gamma ray as it decays to the stable state of ^{46}Ti.

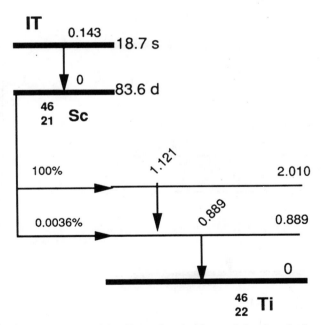

Figure 4.3. Decay scheme for ^{46}Sc. (Reproduced with permission from Lederer and Shirley, 1978.)

The decay of ^{65}Zn is an example of decay by positron emission and electron capture. The decay scheme for ^{65}Zn, produced by the ^{64}Zn(n,γ) reaction is given in Figure 4.4. The decay through the energy level with gamma emission results this time in the stable daughter with one proton less than the precursor, ^{65}Cu. There are three gamma rays emitted with energies of 0.345, 0.771 and 1.115 MeV.

Finally it should be pointed out that the same daughters may be produced from the decay of two different precursors, sharing the same gamma ray transitions in the energy levels of the mutual daughter. An example of this is given in Figure 4.5 for part of the decay schemes of ^{153}Sm and ^{153}Gd. They both have gamma rays produced as the result of transitions between energy levels of ^{153}Eu.

BRANCHING RATIOS

Decay schemes for radionuclides illustrate the mode of decay and the nature of the radiation emitted during decay. They also provide

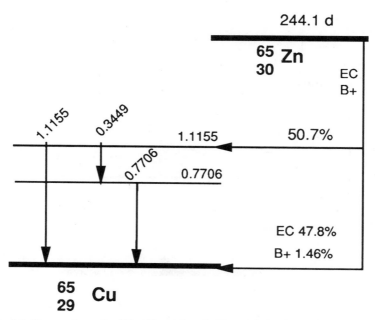

Figure 4.4. Decay scheme for ^{65}Zn. (Reproduced with permission from Lederer and Shirley, 1978.)

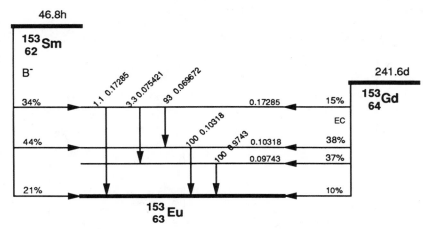

Figure 4.5. Decay scheme for ^{153}Sm and ^{153}Gd. (Reproduced with permission from Lederer and Shirley, 1978.)

quantitative information about the amount of radioactivity that will be produced per disintegration. The proportion of beta and gamma emission are in a fixed ratio dictated by the probability of allowed transitions.

For example the decay scheme for ^{198}Au, in Figure 4.6, shows the disintegration by beta emission to three different energy levels of decay product. Of the disintegrations to the stable daughter product, 0.025% are by β^-, which are emitted with a maximum energy at 1.371 MeV. The majority of disintegrations (98.6%) are with a maximum energy of 0.96 MeV to the 0.4118 MeV energy level of the ^{198}Hg, followed by gamma emission to the ground state. 1.3% of the disintegrations are by emission of β^- with a maximum energy of 0.29 MeV to the 1.0877 MeV energy level of ^{198}Hg. Gamma ray transitions from the 1.0877 MeV level follow one of two paths. Eighteen per cent are to the ground state, with emission of a 1.0877 MeV gamma ray, and 82% to the 0.4118 MeV level, with emission of a 0.6759 MeV gamma ray. All transitions from the 0.4118 MeV level are to the ground state with emission of a 0.4118 MeV gamma ray.

The proportion of gammas produced for a disintegration of a nucleus is called the branching ratio. In the decay of ^{198}Au, for example, it is possible to calculate the gammas per disintegration for the 0.4118 MeV gamma ray in the following way: for every 100 disintegrations there is a probability that 0.025 will go to the ground state. There are 1.3 disintegrations which go to the 1.0877 MeV level, and 18% of the 1.3

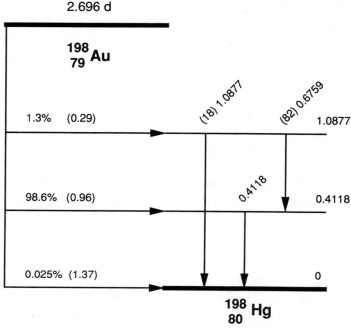

Figure 4.6. Decay scheme for ¹⁹⁸Au. (Reproduced with permission from Lederer and Shirley, 1978.)

(0.23 per 100 disintegrations) will decay to the ground state. The other 82% of the 1.3 (1.06 per 100 disintegrations) decay with emission of a 0.6759 MeV gamma ray to the 0.4118 MeV level. The majority of the β^- transitions (98.6%) go to the 0.4118 MeV too, so for every 100 disintegrations there are 98.6 plus 1.06 from the higher energy levels, a total of 99.67 per 100 disintegrations. Therefore the percentage intensities of the transitions are 99.67 (0.4118 MeV), 1.07 (0.6759 MeV) and 0.23 (1.0877 MeV). A small proportion of the transitions from 0.4118 MeV to the ground state are by emission of conversion electrons. The electrons are emitted from the K and L shells with a ratio of $K/L_1/L_2/L_3$ equal to 687/100/101/45 (Lederer and Shirley, 1978). Since the ratio of K shell electrons to gamma rays e_K/γ is 0.03, if the contribution from the M shells is included the total conversion ratio is just over 0.04 or 4%. The remaining transitions by gamma emission will be just under 96% of 99.67% (the transition intensity) or 95.5%.

Compilations of abundances are found in the literature (Yule and

Grimm, 1981; Erdtmann and Soyka, 1975; Browne and Firestone, 1986) but continue to be updated as new data becomes available. The examples in Table 4.1 are from the most recent comprehensive compilation of values. Care should be taken when referring to the intensity of gamma ray lines since some compilations are in gammas per disintegration, some in per cent abundance, as defined above.

Some compilations take the most abundant gamma ray and normalize everything to it. For example in the case of ^{198}Au, if the values were normalized to the most abundant gamma ray, the intensities would be 100, 0.84 and 0.17 for the 0.4118, 0.6759, and 1.0877 MeV gamma ray energies, respectively. It is interesting to note that in the case of ^{60}Co, shown in Figure 4.7, there are two gamma rays in cascade and there is therefore a total of 200 gammas per 100 disintegrations.

DETECTION OF GAMMA RAYS

This section describes very briefly the main processes involved in the detection of gamma rays. There are several good books available for detailed coverage of the detection and measurement of gamma rays. In particular, an extremely comprehensive book by Knoll (1989) has been revised recently to reflect the latest developments in the technique and a book by Debertin and Helmer (1988) also gives excellent coverage of the topic.

There are three main processes which occur when gamma rays interact with matter: photoelectric absorption, Compton scattering and pair production. The photoelectric effect occurs when a gamma ray interacts with an atom of the absorber and a photoelectron is emitted. The energy of the photoelectron is equal to that of the gamma ray less the binding energy of the photoelectron in its original shell (usually the K shell). The binding energies are typically a few keV for low Z materials to tens of keV for higher Z material. The vacancy in the shell of the absorber atom is filled through capture of a free electron in the absorber material, accompanied by a characteristic X-ray or Auger electron. Normally they are reabsorbed in the material. Photoelectric absorption is the effect which is used to detect and measure the energy of the gamma ray in a semiconductor detector.

The second effect of the interaction of gamma rays with matter is Compton scattering. In this case the gamma ray interacts with an electron in the absorber and some of the energy of the gamma ray is transferred by inelastic scattering to the recoil electron. The result is a gamma ray of some undefined lower energy plus an electron with the residual energy,

Table 4.1. Gamma ray abundances for some commonly used radionuclides

Nuclide	Gamma ray energy, in keV (% intensity)
^{24}Na	1 368.6 (100), 2 754.0 (99.9)
^{27}Mg	170.7 (0.79), 843.8 (73), 1 014.4 (29.1)
^{28}Al	1 779.0 (100)
^{38}Cl	1 642.1 (31.0), 2 167.7 (42)
^{42}K	312.4 (0.35), 1 524.6 (18.8)
^{49}Ca	1 409.0 (0.63), 2 372.0 (0.49), 3 084.5 (92.1), 4 072.0 (7.0)
^{46}Sc	889.3 (99.98), 1 120.5 (99.99)
^{51}Ti	320.1 (93.0), 608.6 (1.18), 928.6 (6.9)
^{52}V	1 333.7 (0.59), 1 434.1 (100), 1 530.7 (0.12)
^{51}Cr	320.1 (9.83)
^{56}Mn	846.8 (98.9), 1 810.8 (27.2), 2 113.2 (14.3)
^{59}Fe	1 099.3 (56.5), 1 291.6 (43.2)
^{60}Co	1 173.2 (99.90), 1 332.5 (99.98)
^{64}Cu	1 345.8 (0.48)
^{65}Zn	1 115.5 (50.75)
^{76}As	559.1 (45), 657.1 (6.2), 1 216.1 (3.42)
^{80}Br	616.9 (6.7), 665.94 (1.08)
^{86}Rb	1 076.7 (8.78)
^{101}Mo	81.0 (3.84), 191.9 (18.5), 506.0 (11.8), 590.9 (16.4), 695.7 (7.2), 1 012.5 (12.8), 1 532.5 (6.0)
^{104m}Rh	51.4 (48.3), 555.8 (0.130)
^{110m}Ag	657.8 (94.6), 763.9 (22.3), 884.7 (72.7), 937.5 (34.4), 1 384.3 (24.3)
^{116m}In	138.4 (3.29), 417.0 (29.2), 818.7 (11.5), 1 097.3 (56.2), 1 293.6 (84.4), 1 507.7 (10.0), 1 752.9 (2.46), 2 112.3 (15.6)
^{124}Sb	602.7 (97.8), 645.9 (7.4), 722.8 (10.9), 1 691.1 (47.1), 2 090.9 (5.49)
^{128}I	442.9 (16.9), 526.6 (1.57)
^{134}Cs	604.7 (97.6), 795.9 (85.4), 802.0 (8.73)
^{140}La	328.8 (20.7), 487.0 (45.9), 815.8 (23.6), 1 596.5 (95.4)
^{141}Ce	145.4 (48.4)
^{152}Eu	121.8 (28.4), 344.3 (26.6), 1 408.0 (20.8)
^{181}Hf	132.9 (35.9), 345.8 (15.1), 482.0 (80.6)
^{187}W	72.0 (10.77), 134.2 (8.56), 479.5 (21.1), 551.5 (4.92), 618.3 (6.07), 685.7 (26.4), 772.9 (3.98)
^{192}Ir	296.0 (28.3), 308.5 (29.3), 316.5 (83.0), 468.0 (47.7), 588.6 (4.47), 604.4 (8.23), 612.5 (5.34)
^{198}Au	411.8 (95.5), 657.9 (0.80), 1 087.7 (0.159)
^{239}U	74.7 (52.5)

Source: Table of Radioactive Isotopes, Browne and Firestone, copyright © 1986, reprinted by permission of John Wiley & Sons, Inc.

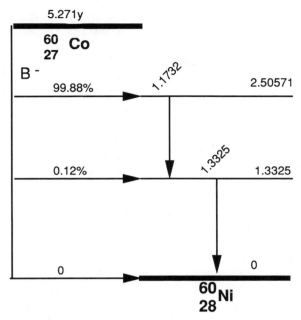

Figure 4.7. Decay scheme for ^{60}Co. (Reproduced with permission from Lederer and Shirley, 1978.)

dependent on the scattering angle, which can range from zero up to a maximum which occurs in the case of a head-on collision. The maximum energy can be calculated from the rest mass energy of the electron (0.511 MeV) and the incident photon energy. The difference in energy between the gamma ray energy and the maximum Compton recoil electron energy tends towards 0.256 MeV when the incident gamma ray energy is large. This Compton effect therefore results in a broad range of gamma ray energies below the energy of the original gamma ray, which will continue to interact by the photoelectric effect or others. The result of this effect is production of a continuous background which can interfere with the detection of gamma rays which undergo the photoelectric effect.

The third interaction is pair production, where the gamma ray loses energy by production of an electron–positron pair. This can only occur if the energy of the gamma ray exceeds 1.02 MeV, which is twice the rest mass energy of an electron (0.511 MeV). All the energy of the gamma ray is lost to the electron–positron pair, so any energy above the 1.02 MeV appears in the form of kinetic energy. The positron will normally annihilate an electron in the absorber, with the production of two annihilation gamma ray photons each with energy of 0.511 MeV. These gamma rays may be absorbed in the material or escape. If the detector is small then both may escape and only the kinetic energy will remain. The result is a photopeak with energy equal to the gamma ray energy minus the 0.511 or 1.02 MeV.

The interaction of gamma rays with matter, described above, may be used as the means to detect them. For a material to be a suitable gamma ray detector it must respond to the gamma ray interactions to produce electrons and also have a means of detecting them. Assuming such a material is available, how will the detector respond to the gamma ray emitter? In the case of photoelectric absorption the gamma ray energy is lost to the photoelectron with a small component producing low energy electrons corresponding to the absorption of the original binding energy of the photoelectron. Provided that the total energy remains in the detector it will be equal to the gamma ray energy. Compton scattering results in a continuum of energy being transferred, equal to the gamma ray energy minus 0.256 MeV and below. There is a certain rounding off of the rise in the continuum at the high energy end due to the binding energy of the electron prior to scattering. There is also the introduction of a slope above the maximum edge.

The pair production process results in the transfer of gamma ray energy to kinetic energy shared by the electron–positron pair. The energy is equal to the gamma ray energy less 1.02 MeV. There is immediately the annihilation of the positron which results in two gamma ray photons of energy 0.511 MeV and they coincide with the original pair production interaction. In a small detector the annihilation photons escape and the deposited energy is the full energy minus 1.022 MeV double escape peak. In a larger detector it may happen that only one escapes and one is absorbed, in which case the energy deposited is 0.511 MeV less than the full photopeak energy.

There are other interactions which may affect the resulting detection of gamma ray photopeaks. If there is β^- emission with the gamma ray emission, then absorption in the material of the detector may result in some secondary radiation in the form of bremsstrahlung. It normally extends to the energy well below the beta emitter. The result is a continuum which is high at the low energy end. Interactions which can occur outside the detector material may also contribute to the spectrum. If a gamma ray is absorbed in material surrounding the detector, the photoelectric effect will result in emission of a characteristic X-ray. So if the detector is surrounded with lead then the characteristic X-ray of lead will be seen. Compton scattering will result in scattered gamma rays with energies below the emitted gamma ray and dependent on the scattering angle but it is always 0.25 MeV or less.

Additional peaks are seen as a result of coincident summing of two gamma rays in cascade, for example in the case of ^{60}Co where the 1.1732 and 1.3325 MeV gammas are emitted almost simultaneously and may be detected as a single gamma of 2.5057 MeV. Figure 4.8 shows the spectrum for ^{24}Na, with the lines and background resulting from all the interactions that may be produced in a detector material, including the gamma rays at 1368.54, 2754.03, 3867.2 and 4238.1 keV, with the corresponding single and double escape peaks.

SEMICONDUCTOR DETECTORS

Unlike charged particles, gamma rays do not create an ionization path as they pass through a material and the detection of gamma rays is dependent on transfer of the energy of the gamma ray to an electron in the absorber. Therefore a gamma ray detector must act as a conversion medium and a detector of the electrons. Semiconductor materials are the

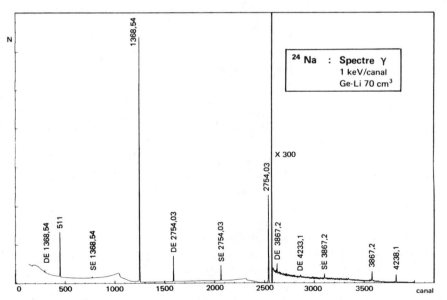

Figure 4.8. The gamma ray spectrum for ^{24}Na shows the peaks due to the four gamma rays plus their corresponding single and double escape peaks, and the annihilation peak at 511 keV. (Reproduced with permission from Legrand et al., 1975.)

best detectors of radiation for this work. Semiconductor detectors have a *p–i–n* diode structure in which the intrinsic region is formed by depletion of charge carriers with application of reverse bias across the diode.

The energy of an electron in a crystalline material is confined to discrete energy bands, separated by forbidden energies. The valence band corresponds to energies of electrons held in specific sites within the crystal. The conduction bands represent the regions where the electrons are free to migrate through the crystal. The two bands are separated by the band gap. In an insulator the band is wide, usually more than 5 eV, but semiconductors have smaller bands, 1.12 eV for silicon and 0.665 eV for germanium. Normally, at temperatures above zero, there is sufficient energy for a valence electron to pass across the gap into the conduction band, leaving a vacancy. This results in the electron–hole pair, the electron in the conduction band and the hole in the valence band. If an electric field is applied across the semiconductor, the electron and the hole will migrate in opposite directions.

In theory a pure semiconductor should have an equal number of electrons in the conduction band and holes in the valence band, which will only be the case in an intrinsic or completely pure semiconductor. In practice it is not possible to obtain sufficiently pure germanium or silicon and the properties will depend on the level of impurity in the material. An impurity in the crystal lattice which has a higher valence than the material will act as a donor and increase the electrons in the conduction band without the corresponding holes in the valence band. The conduction of this *n*-type material is almost totally dependent on the flow of electrons, and the holes play a small role. The addition of an impurity with a lower valence will result in a shortage of electrons in the conduction band. If an electron migrates to fill the gap, it will be less firmly attached than by a typical valence bond so electron sites can be created in the forbidden band. The result will be holes in the valence band which exceed the electrons in the conduction band. This is called a *p*-type material.

The most practical configuration for a semiconductor detector is produced on the junction between *n*- and *p*-type materials. At the junction, where there is a sharp gradient, there is diffusion of conduction electrons into the *p*-type material where they combine with holes. Similarly holes will diffuse into the *n*-type material. At equilibrium the space at the junction becomes a depletion region where the numbers of donors and acceptors are equal. This layer exhibits a high resistivity. When voltage is applied in the reverse polarity, it produces an electric field across the neutral region. The positive holes in the *n*-type material are attracted to the electrode on the *n* side and the electrons are attracted to the electrode on the p side. The net result is a dead space between, causing a high resistance to the passage of current. As the bias is increased so the thickness of the depletion layer increases until it reaches the breakdown voltage of the junction. The depletion depth is inversely proportional to net electrical impurity concentration and extremely pure material is required to obtain large sensitive volumes. Small currents can operate if the minority carriers are attracted across the junction. This constitutes the small leakage current which is present in a detector. There will also be some thermal generation of electron–hole pairs in the depletion region, which is minimized by cooling the material. At 77K or $-200°C$ reverse leakage currents are in the range 10^{-9} to 10^{-12} A.

In order for a semiconductor material to be used as a radiation detector the electrical charge created by the ionization energy must be collected. Several thousand volts are applied to collect the charge carriers efficiently. A gamma ray interacting in the material of the detector produces free electron–hole pairs. If electron–hole pairs are created in the compensated region the free electrons will move towards the positive contact and the free holes in an opposite direction, thus producing an electrical signal. The signal produced by a gamma ray is proportional to the gamma ray energy (approximately 0.05 pC MeV^{-1}). This signal is a combination of many ($>300,000$ MeV^{-1}) electron–hole pairs and therefore statistical fluctuations are relatively small. The resolution of the detector depends on this fluctuation, which can be predicted with Poisson statistics. The resolutions obtained with semiconductors are in fact better than predicted and so the Fano factor is used to define the statistical fluctuation in charge collection. The Fano factor is the ratio of the observed variance to the Poisson predicted variance. The Fano factor is much less than unity for semiconductor material and the value is as low as 0.13 for germanium.

Germanium of the p-type has an impurity level of the order of 10^{16} atoms per m^3. This means that the germanium needed to make a high purity germanium detector can tolerate one impure atom for every 1.5×10^{12} germanium atoms. Only recently has this been achieved. Until the mid-1970s the purity required to produce large-volume detectors could only be attained by doping p-type germanium crystals with n-type impurity, lithium, in a process called lithium drifting. Germanium crystals of sufficient purity are now grown routinely and lithium-drifted detectors are no longer produced in quantity. High purity germanium detectors may be p-type, which were the first produced, or n-type. Both forms of detector are represented diagrammatically in Figure 4.9.

The p-type detector consists of a cylinder of pure germanium so that collection at the electrode is through a constant layer of germanium. The

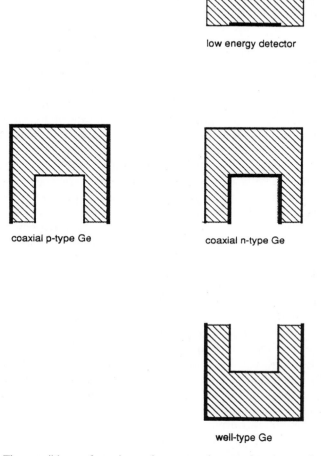

Figure 4.9. The possible configurations of a germanium semiconductor detector. The thickened line indicates the location of the "dead layer".

outer electrical contact is normally lithium which is diffused into the germanium to a depth of about 0.5 mm. This forms a dead layer through which the radiation must pass before it is detected. The inner (in the hole) electrical contact is made by evaporating a metal such as gold onto the semiconductor surface. A more robust form of contact is attained through ion implantation of boron which buries the contact below the surface of the germanium. In the *n*-type detector the configuration is reversed and the inner contact is made by diffused lithium. The outer contact is made by ion implantation which makes the entrance window much thinner (0.3 μm). This type of detector is therefore better than the *p*-type for the detection of low energy gamma rays.

Planar germanium detectors are also cylindrical but are thinner with a depth of only 5–25 mm. They consist of a disk of germanium with a lithium-diffused contact on one side and a surface barrier or ion-implanted *p*-type contact on the other. The thin *p*-type contact would be on the front face of the detector. They are used to measure low energy gamma rays with the advantage that being thin they do not stop the more energetic high energy gamma rays, reducing the interference effects.

REFERENCES

Amiel, S. (1981), "Neutron counting in activation analysis" in S. Amiel (ed.) *Nondestructive Activation Analysis*, Studies in Analytical Chemistry 3, Elsevier, Amsterdam, pp. 43–52.

Browne, E. and R. B. Firestone (1986), *Table of Radioactive Isotopes,* Wiley, New York.

Debertin, K. and R. G. Helmer (1988), *Gamma- and X-ray Spectrometry with Semiconductor Detectors*, North Holland, Amsterdam.

Erdtmann, G. and W. Soyka (1975), "The gamma-ray lines of radionuclides ordered by atomic and mass number. Part I. $z=2$–57 (helium–lanthanum)," *J. Radioanal. Chem.*, **26**, 375–495. "Part II. $z=58$–100 (cerium–fermium)," *J. Radioanal. Chem.*, **27**, 137–286.

Knoll, G. F. (1989), *Radiation Detection and Measurement*, Wiley, New York, 2nd Ed.

Lederer, C. M. and V. S. Shirley (eds) (1978), *Table of Isotopes*, Wiley, New York, 7th Ed.

Legrand, J., J. P. Perolat, F. Lagoutine, and Y. Le Gallic (1975), *Table de Radionuclides*, Commissariat a l'Energie Atomique.

Mantel, M. and S. Amiel (1981), "X-ray spectrometry" in S. Amiel (ed.), *Nondestructive Activation Analysis*, Studies in Analytical Chemistry 3, Elsevier, Amsterdam, pp. 25–42.

Peisach, M. (1981), "Prompt techniques" in S. Amiel (ed.), *Nondestructive Activation Analysis*, Studies in Analytical Chemistry 3, Elsevier, Amsterdam, pp. 93–111.

Yule, H. P. and C. A. Grimm (1981), "Tables of gamma-ray peaks observed in reactor activation analysis" in S. Amiel (ed.), *Nondestructive Activation Analysis*, Studies in Analytical Chemistry 3, Elsevier, Amsterdam, pp. 321–360.

CHAPTER
5
SPECTROMETRY EQUIPMENT

The gamma ray counting system for multielemental trace element analysis normally consists of a semiconductor detector and its associated electronics to collect and shape the pulses produced. The pulses are sorted in spectrum analyzers which are currently based on computers for storage and processing in memory or on disk. Computer programs are used to find peaks, calculate count rates and correct for dead time and other losses due to the electronics. Together with computers, mechanical sample changers can be used to automate the entire counting procedure. Detailed descriptions of gamma ray detectors, associated electronics and analyzers with computer processing are found in excellent publications by Debertin and Helmer (1988) and Knoll (1989).

DETECTOR

The construction of all semiconductor detectors is very similar, a typical detector configuration is shown in Figure 5.1. The shape and size of the semiconductor crystal will depend on whether the detector is n- or p-type and whether it is coaxial, planar or well-type. However, they are all housed in an aluminum casing, they are all under vacuum and they all operate at liquid nitrogen temperature. The detector is cooled during operation, usually via a copper rod dipstick in contact with a dewar of liquid nitrogen. The lithium-drifted germanium detectors must always remain in contact with liquid nitrogen otherwise the lithium drifts out of the crystal. The pure germanium crystals do not deteriorate at room temperature and can be transported and stored without liquid nitrogen.

There is a choice of configuration for the detector and the position of the detector in the dewar will depend on the geometry required for counting the sample. Commonly a sample will be sited above the end-cap of a vertical dipstick detector but it may be more convenient to use a horizontal configuration if, for example, the sample is being delivered via a vertical pneumatic tube. It is possible to get any angle on the head, although it is usual to operate a vertical or horizontal dipstick model.

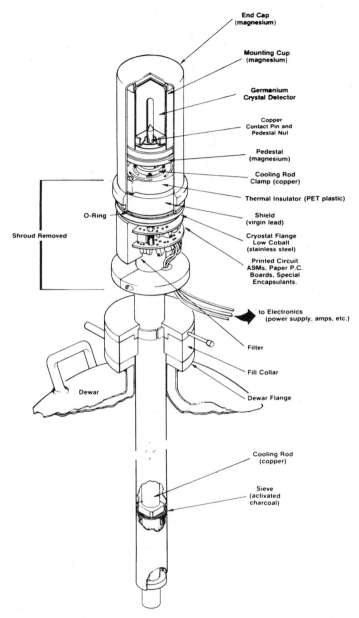

Figure 5.1. Diagram of a high purity germanium semiconductor detector, showing the crystal mounted in a horizontal configuration. (Reprinted with permission from EG & G ORTEC, Oak Ridge.)

The dipstick may sit in the neck of the dewar or be mounted in the base as an integral arrangement. There are vertical and horizontal dipsticks; sidelooking, downlooking and J-type buckets and portable devices for field measurements; the various arrangements are shown in Figure 5.2. Detectors are available where the head can be removed from the cryostat for transportation or replacement in a new configuration. It is now possible to get electric cooling devices which do not require liquid nitrogen and these are particularly useful for fieldwork.

The casing of the detector is normally made of aluminum, allowing the passage of gamma rays with the minimum attenuation. However, there is significant loss of low energy gamma ray and X-ray energy in the aluminum and to overcome this beryllium windows are fitted to allow the passage of low energy radiation. Detectors with the n-type configuration are preferred for low energy work because the ion-implanted electrode has a much thinner dead layer than the diffused layer of the p-type detector. The form of the n-type detectors is shown in Figure 5.3. The planar detector is designed specifically for X-ray and low energy gamma ray analyses. These detectors are thin disks of semiconductor material, with the ion-implanted electrode facing the thin beryllium

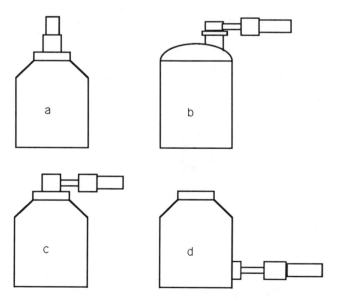

Figure 5.2. The dewar configurations for germanium detectors include (a) vertical mounting, and three types of horizontal mounting: (b) offset, (c) central, and (d) integral.

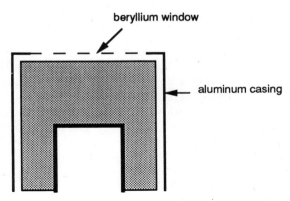

Figure 5.3. The *n*-type detector has a thin ion-implanted outer contact and a beryllium window in the aluminum casing to allow the detection of low energy X-rays and gamma rays.

window, as shown in Figure 5.4. High energy gamma rays will pass through the crystal without being detected so there is a limited energy range. The resolution of these detectors for low energy gamma and X-rays are superior if optimized.

The efficiency of the detector for the detection of gamma rays is dependent on the size of the active volume of the crystal. The efficiency is usually quoted in the specifications for the detector measured at the 1.33 MeV peak of ^{60}Co at a distance of 250 mm from the detector. The value is compared to the efficiency of a 76 mm × 76 mm sodium iodide detector for the 1.33 MeV ^{60}Co peak measured at 250 mm, which is 1.2×10^{-3}. Germanium crystals can now be made very large and consequently the efficiency of such detectors is high. For neutron

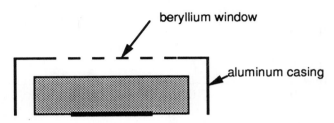

Figure 5.4. The planar germanium crystal for detection of low energy X-rays and gamma rays has the ion-implanted contact at the top surface and a thin beryllium window in the casing.

SPECTROMETRY EQUIPMENT

activation analysis a detector of about 0.04 dm³ volume with an efficiency of 20% would be usual, but it is possible to obtain crystals with efficiencies of up to 100%. The planar detectors which are designed for low energy work are not very efficient at the higher energy end but a thin beryllium window allowing through the low energies will give higher efficiencies than the coaxial detector at the lower energies. Similarly the *n*-type detector with a beryllium window will give good efficiencies throughout the gamma ray energy range. Figure 5.5 gives a comparison of the efficiency curves for the detectors to show the relative advantages of the different configurations. The maximum efficiency for a germanium detector will be obtained with a well-type detector. The sample can be placed inside the crystal which is drilled out to accommodate it. This gives almost complete collection of the radiation and very high efficiencies.

The bias applied across the crystal is supplied through the preamplifier by a separate high voltage unit. The bias supply may be switched to positive or negative and will depend on the form of the crystal. The size of the bias supply is usually 0–5 kV and the applied bias will depend on the optimum operating voltage of the crystal. This operating voltage, which will be of the order of a few thousand volts, is measured and recommended by the manufacturer. Most preamplifiers are designed to protect the detector electronics against a sudden surge in power but care

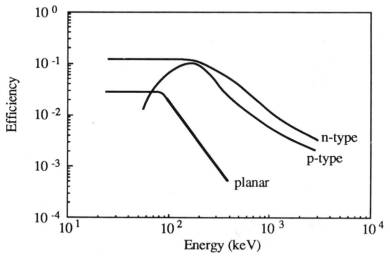

Figure 5.5. The efficiency curves for a coaxial *p*-type, a coaxial *n*-type and a planar germanium detector. The planar and *n*-type detectors, with thin windows, have high efficiencies at low energies. The thin planar detector has poor efficiency at high energies.

should be taken to apply the bias gradually. The bias supply should not be applied unless the crystal is cooled in liquid nitrogen and some preamplifiers have protection so that if the temperature of the crystal rises the bias is automatically switched off.

It is usually necessary to include some lead shielding round the counting position to reduce the background counts due to naturally occurring gamma ray emitters and the other samples waiting to be counted, if they are close by. Lead shielding configurations suitable for a germanium detector are available commercially, commonly in the form of a hollow cylinder with a sliding lid for access.

The pulses produced in the detector are very small and so the charge sensitive preamplifier, usually a field-effect transistor (FET), is mounted as close to the detector as possible. Normally this preamplifier forms part of the detector configuration. The FET can be cooled for minimum noise particularly with planar detectors, in which case the preamplifier is an integral part of the cooling system. It is quite usual to have preamplifiers which are cylindrical and form part of the dipstick arrangement.

PREAMPLIFIER

The very small pulse which is produced in the semiconductor detector, is amplified with a gain of about 2 V pC^{-1} to approximately 0.1 V MeV^{-1} of gamma energy in the charge-sensitive preamplifier. This signal is sent to the shaping amplifier where it is shaped and amplified, usually up to a maximum of 10 V, depending on the gain setting. The resulting pulse, which still has an amplitude corresponding to the original energy of the gamma ray, is sorted by a pulse height analyzer. A schematic diagram of the electronics is given in Figure 5.6.

The preamplifier is usually closely associated with the detector forming an integral part of the cryostat system to avoid electronic noise. If very low noise conditions are required the FET itself is cooled in the cryostat system. There are really only two types of electronics associated with preamplifiers for detectors for gamma ray analysis. These are charge-sensitive preamplifiers that employ either dynamic charge restoration (*RC* feedback) or pulsed charge restoration (pulsed optical feedback or transistor reset) methods to discharge the transistor.

In the *RC* feedback preamplifier, the feedback resistor is connected in parallel with the FET, so that each pulse decays with a time constant. For most applications the noise level of these preamplifiers is low enough not to affect the resolution of the system. However, in the case of low energy counting it can become a problem.

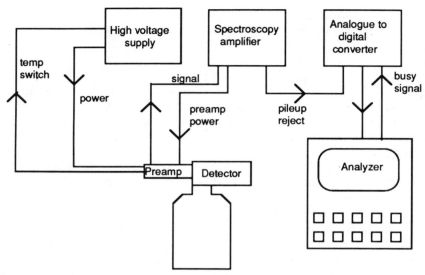

Figure 5.6. Schematic diagram of a modern gamma ray spectrometry system showing the electronics required for quantitative multielement analysis.

Pulsed optical reset preamplifiers were developed to overcome the noise problem by replacing the feedback resistor with a light-emitting diode. This type of preamplifier is used widely on low energy detectors such as planar detectors, where resolution is important. When a photon produces a charge in the crystal a voltage step is produced in the preamplifier. To prevent saturation of the preamplifier, when the output level reaches a certain voltage a short pulse is sent to a light-emitting diode producing a light pulse which discharges the FET. The noise is decreased but the recovery times and dead times will be extended as a result. The pulsed optical reset preamplifier is therefore not used on coaxial detectors.

The transistor reset preamplifier was developed to overcome the problem of the higher dead times that occur when processing pulses at high count rates. In this case the feedback capacitor is discharged by means of a transistor switch connected to the FET gate. The transistor adds some noise but this is usually acceptable in return for the high count rates that can be used. The main application of transistor reset preamplifiers is in situations where high count rates are processed. If count rates become too high for the preamplifier to cope with there will be pileup of pulses occurring. Some preamplifiers now may have a visual warning such as a red light to indicate that the count rate is too high for the preamplifier circuitry.

SPECTROSCOPY AMPLIFIER

The amplifier increases the size of the small signal from the output of the preamplifier to a pulse of amplitude up to about 10 V. An output pulse from the amplifier will have an amplitude in the range of 0–10 V, dependent on the gain setting. Normally there will be a coarse gain and a fine control for very precise setting of the amplifier gain.

The second purpose of the amplifier is to improve the signal to noise ratio of the signal pulses. Semi-Gaussian shaping is generally used, with differentiation and integration of the signal. The time constants of these two processes can be adjusted independently. They are kept longer than the charge collection time in the detector but they are kept as short as possible to avoid pileup of pulses in the amplifier. Normal shaping constants are about 1–6 μs, although tenths of microseconds are used for applications with very fast count rates.

At the end of the pulse from the preamplifier output there will be a slight undershoot of the pulse below the baseline, which lasts for some time and is therefore likely to affect the processing of the next pulse. The undershoot is removed by adjusting a resistance value in the so called pole–zero cancellation circuit. The correction is made by inspecting the output pulses from the amplifier on an oscilloscope and adjusting the pole–zero setting to eliminate the undershoot.

The amplitude of the pulse is the important information carried by the pulse. Therefore the amplitude measured relative to a zero baseline will be badly affected by any shift in baseline level. If there are a number of pulses and the baseline is not restored to zero, then there will be a detrimental effect on the spectrum. Baseline shifts are not a problem if bipolar pulses are used but the signal to noise ratio tends to be superior with unipolar pulses. Most amplifiers have automatic baseline restoration circuits in them to restore the baseline level after the pulse.

Spectroscopy amplifiers can work with quite high count rates before pulse pileup begins to occur. However, in the case of high count rates it is possible for pulses to overlap and to be seen as one peak with the energy of the sum of the energies of the pulses. Amplifiers can be obtained with pileup rejection circuits built into them. The circuit will reject any pulses which are seen as being added together and a signal from the amplifier will tell the analyzer to ignore that pulse. The time lost due to the pileup rejection circuit must be included in the dead time of the counting system.

When very high count rates are to be handled a gated integrator may be used to eliminate charge collection time effects which would seriously distort the spectrum from a conventional semi-Gaussian filter amplifier. The charge collection effect is significant for large-volume detectors when the pulses from high energy gamma rays have the same total charge but different rise times in the detector. The distortion is most significant when short shaping times are used. The gated integrator output is obtained by integration of the entire unipolar signal resulting in the elimination of the charge collection time effects even at shaping times in the order of 0.25 μs. Consequently when short shaping times are used, to increase the throughput and reduce pulse pileup effects, the energy resolution is affected to a lesser extent than with a conventional amplifier.

PULSE HEIGHT ANALYZER

The output from the main amplifier is a peak of nearly Gaussian shape with an amplitude proportional to the gamma ray energy which enters the detector. The analogue–digital converter (ADC) changes this pulse into a digital signal proportional to the pulse height which is deposited as a count in the appropriate channel number of the analyzer. The ADC may be a separate module which is placed between the amplifier and the analyzer input. However it is now just as likely to be incorporated into the analyzer itself, in which case it could be a board inside a computer. The amplitude of the output pulse from the amplifier will be in the range 0–10 V. The conversion gain of the ADC specifies the number of channels over which the amplitude range will be spread. It would be normal to use a conversion gain of 2048, 4096 or 8192 for multielement gamma ray spectrometry.

Traditionally an ADC with a Wilkinson type ramp converter is used. The Wilkinson ADC digitizes a pulse by charging a capacitor to the amplitude of the input pulse and then discharging the capacitor at a constant rate. A crystal-controlled pulse timer is counted in a register during the discharge time and the total number of pulses produced results in the event address, which is proportional to the pulse amplitude. The time for the ADC to process a pulse increases linearly with the channel number of the pulse that is being processed. A typical clock time of 100 MHz results in an ADC processing time of 10 ns per channel. Once the pulse has been digitized there are a few additional microseconds to store the pulse. Therefore the total processing time is $A/v + B$, where A is the channel number, v is the clock frequency and B is the pulse storage time.

An alternative type of ADC is the successive approximation system. A series of comparators are used to successively re-estimate the pulse amplitude to the required precision. The successive approximation ADC is inherently faster than the Wilkinson system, with processing times between <10 and <25 μs, since the process time increases as the logarithm of the number of channels. The main disadvantage is that linearity is often poorer.

The processing time which is lost during counting is called the dead time. The dead time is the time that the ADC is busy while processing one pulse and cannot accept another. This loss of pulses can be corrected for by measuring the time for which the ADC is busy. The analyzer interrogates the ADC busy signal at a very high frequency while incrementing a live time clock. The live time clock only operates when the ADC is operating. The difference between the time incremented on the live-time clock and the real-time clock in the analyzer is the dead time. The dead time of the system, in terms of a percentage of the real time, is indicated on a meter on the front panel of an ADC unit or on a display screen. Provided that the live time is recorded during the counting period it is simple to present the data as counts per live second to correct for any losses.

An alternative means of correcting for lost pulses is using the virtual pulse generator which is based on the technique developed by Westphal (1987). This technique provides accurate real-time correction of counting losses at high count rates, dynamically adding the fractional counting losses to the spectrum as they occur rather than extending the measurement duration as in live-time correction.

The role of the analyzer in pulse processing may be seen simply as a data acquisition system, to collect the counts in the channel address which corresponds to the appropriate gamma ray energy. The resulting gamma ray spectra are usually displayed on a screen or as an x–y plot. Nowadays most analyzers are based on computer systems, usually micros and personal computers, and the number of channels are only limited by the memory size. Each channel represents one small equal energy range, for example if each channel represents 1 keV energy, an analyzer with 4096 channels would cover the energy range from 0–4096 keV. Channel 2000 would therefore represent the gamma rays with energies between 1999 and 2000 keV. Each count corresponding to a pulse originating from a gamma ray with an energy between 1999 and 2000 keV will be added to channel number 2000. The gamma ray spectrum may be considered simply as a plot of the content of each channel against channel number, as shown in Figure 5.7.

The counting time for the analysis is usually controlled by the analyzer but it is possible for the timing to be controlled by an external clock which stops and starts the analyzer. The external control is useful where the sample is counted soon after irradiation and the start of analysis has to be coordinated with the end of irradiation which is controlled by an external clock.

GAMMA RAY SPECTRUM

The gamma ray spectrum is characteristic for each gamma ray emitter and there are books available showing semiconductor gamma ray spectra for all the radionuclides (Heath, 1974; Legrand et al., 1975). The shape of a gamma ray spectrum is due to the interactions that take place in the detector. As an example, the gamma ray spectrum for ^{60}Co consisting of two gamma rays at 1173.22 and 1332.51 keV is shown in Figure 5.8. The large peaks are seen at the characteristic energies of the gamma rays, as a result of the photoelectric effect. In addition there is low energy

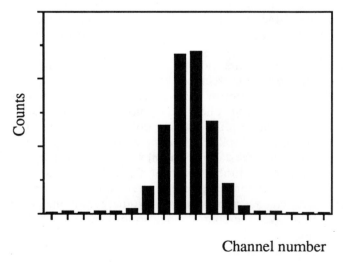

Figure 5.7. A gamma ray peak is represented in a pulse height analyzer as a plot of the counts in a channel against the channel number.

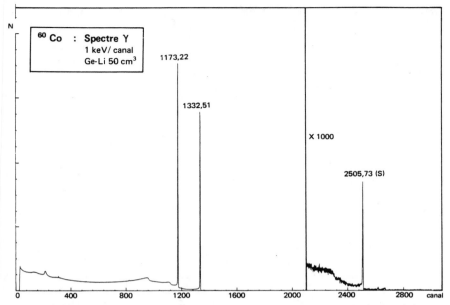

Figure 5.8. The gamma ray spectrum for ⁶⁰Co, showing the low energy background due to the Compton effect and the single and double escape peaks due to pair production. (Reprinted with permission from Legrand et al., 1975.)

background due to the Compton effect and the peak due to coincident summing of the two gamma rays is seen at 2505.73 keV.

If the electronics are operating normally the relationship between the analyzer channel number and the corresponding gamma ray energy should be linear. Therefore the plot of gamma ray energy against channel number should be a straight line with a slope dependent on the amplifier gain, which may or may not go through zero depending on the way that the ADC is set up. The usual method of calibrating the system is to use radionuclides with well-known gamma ray energies, such as commercial reference sources of ⁶⁰Co, ¹³³Ba and ¹³⁷Cs, and those listed in Table 5.1 provide a useful range of energies. The sources are placed on the detector and the channel numbers noted for the maximum number of counts in the peak. If the known energy of the gamma ray line is plotted against channel number the result should be a straight line. The energy calibration is defined by the slope (keV per channel) and the offset (keV).

As the pulses are counted into the appropriate address in the analyzer, the spectrum is built up for the whole energy range and the spectrum can be displayed on a CRT or video monitor. The resolution given in

Table 5.1. Useful reference sources for efficiency calibration

Source	Half-life	Energy (keV)	Intensity (%)
^{22}Na	2.602 y	1 274.5	99.94
^{24}Na	14.659 h	1 368.6	99.994
		2 754.0	99.881
^{46}Sc	83.83 d	889.3	99.984
		1 120.5	99.987
^{54}Mn	312.2 d	834.8	99.975
^{57}Co	271.77 d	14.4	9.3
		122.1	85.63
		136.5	10.58
^{60}Co	5.271 y	1 173.2	99.89
		1 332.5	99.983
^{65}Zn	244.1 d	1 115.5	50.65
^{88}Y	106.61 d	898.0	94.2
		1 836.0	99.30
^{109}Cd	1.267 y	88.0	3.68
^{110}Ag	249.76 d	446.8	3.72
		657.8	94.4
		677.6	10.40
		687.0	6.44
		706.7	16.6
		744.3	4.70
		763.9	22.39
		818.0	7.32
		884.7	72.7
		937.5	34.31
		1 384.3	24.25
		1 475.8	3.99
		1 505.0	13.04
^{124}Sb	60.20 d	602.7	98.0
		645.9	7.3
		722.8	11.3
		1 691.0	48.5
		2 090.9	5.66
^{133}Ba	10.54 y	53.2	2.19
		79.6	2.62
		81.0	34.1
		276.4	7.16
		302.9	18.31
		356.0	62.00
		383.9	8.92

Table 5.1. Continued

Source	Half-life	Energy (keV)	Intensity (%)
^{134}Cs	2.062 y	604.7	97.63
		795.8	85.52
^{137}Cs	30.0 y	661.7	85.2
^{152}Eu	13.33 y	121.8	28.40
		244.7	7.51
		344.3	26.58
		411.1	2.23
		444.0	3.12
		778.9	12.96
		964.1	14.62
		1 085.9	10.14
		1 112.1	13.54
^{182}Ta	115.0 d	100.1	14.23
		152.5	7.02
		222.1	7.57
		1 121.3	35.3
		1 189.0	16.42
		1 221.4	27.20
		1 231.0	11.57
^{192}Ir	73.83 d	296.0	28.7
		308.5	29.8
		316.5	83.0
		468.1	47.4
		588.6	4.49
		604.4	8.11
		612.5	5.28
^{203}Hg	46.60 d	279.2	81.50
^{207}Bi	32 y	569.7	97.9
		1 063.7	74.1
		1 770.2	6.87
^{241}Am	432.7 y	26.3	2.41
		59.5	35.9

Source: Vaninbroukx, 1985

the specification for a detector is a measure of the peak width produced by the crystal. Usually the resolution is quoted for the 1332 keV gamma ray energy of ^{60}Co, and is the full width at half the maximum height (FWHM) of the peak. The energy resolution is dependent on the charge collection, a characteristic of the detector material itself but there is also the possibility of line broadening due to the electronics of the counting system beyond the detector. Degradation of the spectrum can be caused by noise on the baseline of the preamplifier signal or by pulse pileup which results in additional background counts due to summed peaks.

The best resolution theoretically obtainable with a germanium crystal at 1.33 MeV, calculated using the Fano factor, is 1.645 keV. For activation analysis it is usual to use a detector with a FWHM of 1.7–2.1 keV, as a rough indication of acceptable resolution. A planar detector does not operate at such a high energy and the FWHM is usually given for the 5 keV X-ray and the 122 keV gamma ray of ^{57}Co. The resolution of the planar detector is better at the low energy end of the spectrum than for cylindrical crystals, with a typical resolution at 122 keV of 550 eV. The best large-volume crystal in comparison would be an n-type hyperpure germanium detector which would have a FWHM of about 700 eV at best. Resolution is possibly less critical in multielemental analysis than it was in the past now that the software available for spectral evaluation can separate overlapping peaks successfully.

The ratio of the peak height of the gamma ray to the height of the Compton edge is also a function of the performance of the detector and the peak/Compton ratio is quoted in the specification for a detector. The value given is usually the height of a ^{60}Co peak at 1332 keV compared to the average Compton plateau between 1040 and 1096 keV. The value of the peak/Compton ratio will depend on the resolution of the detector and also the proportion of events which are found in the full energy peak. The higher the ratio the better and for a coaxial detector the value is normally in the range of 40–50.

DATA PROCESSING

These days the analyzer not only carries out pulse height analysis and controls the timing of the analysis, it is also the data processing system. Since modern analyzers are based on computers they are capable of carrying out the whole process of analyzing the data. The data processing consists of a number of stages including storage of spectra, peak search, nuclide identification, peak area determination, decay and efficiency corrections.

Until recently the analyzer systems were based on microprocessors. The stand-alone type provides a purpose-made analyzer for spectrometry with software designed for the system. The operating system on such an analyzer is not readily accessible and it can be cumbersome to transfer data from outside the system, or more importantly, from the system into another computer. The main advantage of such a system is that the programs are good and the analyzers are designed to allow several users to operate with a variety of tasks simultaneously. Systems that can be interfaced with computers are more flexible because they can be used as a stand-alone analyzer as well as having the capability to be connected to a separate computer if and when required. The main advantage of this is the access to routine software for other operations, such as statistical packages, spreadsheets, graphics and chemical data management programs. The main disadvantage of such a system is that it is more difficult to inspect the spectrum once it has been transferred from the analyzer to the computer.

The personal computer is an inexpensive alternative to the larger computer based systems used in the past. Because they are standard systems the analyzer can be used with other programs for data handling and presentation. There are several systems available from different manufacturers, depending on the combination of the ADC, analyzer buffer and computer. The ADC and the analyzer buffer may be separate modules or cards inside the computer. They may be separate or combined in one unit. The analyzer buffer includes its own microprocessor and memory so that it operates quite independently of the computer during counting, leaving it free to run other programs. The addition of an input/output board to the system will enable the control of other operations such as sample changers.

It is now possible to buy a personal computer which will do all the jobs that were done by larger computer systems. In gamma ray spectrometry this has had quite a significant effect on the choice of system. In the past the cost of an analyzer was usually greater than the cost of a detector to go with it. Consequently there was an advantage in buying an analyzer that could support many detectors running simultaneously. With the introduction of personal computers, at a time when hyperpure germanium detectors were introduced making lithium-drifted germanium detectors obsolete, the balance has changed and it is possible to buy an analyzer which is relatively cheap compared to the cost of a good detector. The result is a tendency away from multiuser systems and a trend towards analyzers operating with a single detector.

The trend towards simplification of the spectrometry system continues with the recent introduction of high voltage supply, amplifier and ADC combined in a single module to form the connection between detector preamplifier and personal computer. These modules are fully computer controlled and include facilities for automatic calibration, pole–zero, digital stabilization and sample changer control. The complete systems

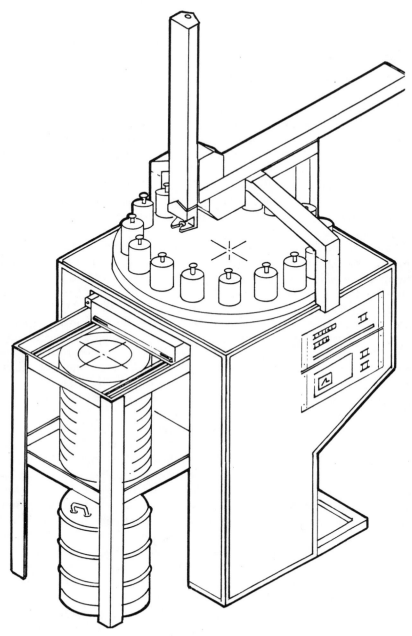

Figure 5.9. A carousel type sample changer for large samples, designed for a standard lead castle. (Reprinted with permission from EG & G Instruments, Wokingham; L. E. Pink Engineering, Ltd., Reading; and Fab Cast Engineering Ltd., Dartford.)

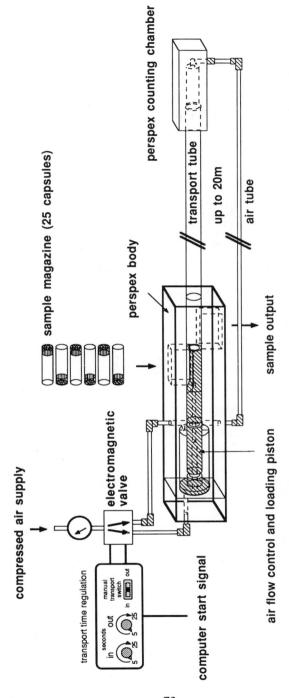

Figure 5.10. A pneumatic sample changer for up to 25 samples. The sample is transferred to the counting position with compressed air. (Reprinted with permission from M. Bichler, Atominstitut der Österreichischen Universitäten, Vienna.)

73

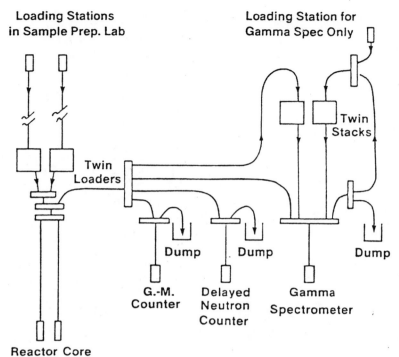

**Loading Stations
in Sample Prep. Lab**

**Loading Station for
Gamma Spec Only**

Twin
Loaders

Twin
Stacks

Dump   Dump   Dump

G.-M.   Delayed   Gamma
Counter   Neutron   Spectrometer
            Counter

Reactor Core

Figure 5.11. A completely automated system, based on pneumatic transfer, for neutron activation analysis with gamma ray spectrometry and delayed neutron counting. (Reprinted with the permission of AECL Research, Canada.)

cost less than a good hyperpure germanium detector and the software programs used in personal computers are equal to those used on the larger computers in the past.

Basically all systems will now carry out a peak search to locate the gamma rays and their energy. The peak area will be evaluated and the data presented as counts per second or activity. The system will attempt to identify the source of the gamma ray energy. The counts may be corrected for detector efficiency and a calculation of the activity made with correction for decay.

AUTOMATION

It is a simple task to program computer based analyzers to carry out all the analytical processes in a totally automatic way. All that is required is a simple program stringing together the steps of the process: counting the sample, storing the spectrum and finally processing the data. These processes can be repeated for a number of samples in sequence. The samples can be changed over manually between counts or an automatic sample changer can be used. This involves the use of a mechanical or pneumatic sample changer to move the appropriate sample into position over the detector plus an operating system, hardware or software controlled, to initiate the sample changing procedure.

Only recently have manufacturers begun to appreciate the market for mechanical sample changers, mainly due to the increased interest in environmental counting. Consequently there is a wide range of models that have been made in house to suit the particular needs of the laboratory. There are carousel types that normally operate in a horizontal plane for use with a vertical dipstick type detector, as shown in Figure 5.9. There are pneumatic systems available commercially, for transfer of samples to and from the detector, shown in Figure 5.10. A sample changer has also been developed which is based on an articulated robot arm (Thompson et al., 1988).

Control of the sample changer can be very simple, for example a logic signal can be used to indicate that the analyzer has stopped counting or that the ADC is switched off. Alternatively, a computer based analyzer can control the sample changer directly as part of the analysis program. Both sample changer and analyzer can be controlled by a microcomputer and delivery of the sample for counting may be part of a totally automated irradiation and counting system such as is used for short irradiations. This sort of system, shown in Figure 5.11, is available commercially.

REFERENCES

Debertin, K. and R. G. Helmer (1988), *Gamma- and X-ray Spectrometry with Semiconductor Detectors*, North Holland, Amsterdam.

Heath, R. L. (1974), *Gamma-ray Spectrum Catalogue; Ge(Li) and Si(Li) Spectrometry Volume 2, 3rd Ed.*, Aerojet Nuclear Company, United States Energy Research and Development Administration, ANCR-1000–2 (Available from NTIS).

Knoll, G. F. (1989), *Radiation Detection and Measurement*, Wiley, New York, 2nd Ed.

Legrand, J., J. P. Perolat, F. Lagoutine, and Y. Le Gallic (1975), *Table de Radionuclides*, Commissariat a l'Energie Atomique.

Thompson, C. M., A. Sebesta, and W. D. Ehmann (1988), "A robotic sample changer for a radiochemistry laboratory," *J. Radioanal. Nucl. Chem.*, **124**(2), 449–455.

Vaninbroukx, R. (1985), "Emission probabilities of selected gamma rays for radionuclides used as detector calibration standards," presented at the Advisory Group Meeting of the International Atomic Energy Agency, IAEA-TECDOC-335, IAEA, Vienna.

Westphal, G. P., Th. Kasa, and W. Roch (1987), "Trends in instrumentation for activation analysis of short-lived nuclides," *J. Radioanal. Nucl. Chem.*, **110**(1), 9–31.

CHAPTER

6

ACTIVATION SPECTROMETRY

Gamma ray spectrometry provides the most valuable technique of analysis for samples activated with neutrons. Identification of the gamma ray lines provides information about the elements in the sample. Peak area evaluation to determine the activity of the nuclide can be used as a quantitative method of determining the element in a sample. The evaluation can be made in an empirical way using physical constants and measured fluxes or by direct comparison with a standard containing a known amount of the element of interest.

NUCLIDE IDENTIFICATION

Gamma ray spectrometry can be used to identify gamma ray energies and consequently the radioactive species which are producing them. In the case of a sample which has been irradiated, the identification of the radionuclide that produces the gamma ray will be used to identify the stable target nucleus in the sample and hence the element.

Tables of gamma ray energies are available for identification of unknown lines, listed in order of increasing energy and sometimes grouped according to half-lives for ease of identification (Erdtmann and Soyka, 1975). If the spectrum is of a completely unknown sample then each line must be examined for possible sources using tables of gamma ray energies. There may be more than one possible source for a particular gamma ray energy, in which case the first check is to see if any other lines from the same radionuclide are also present in the spectrum, and whether they are in the expected ratios, taking into account the relative intensities of the gamma rays and the efficiency of the detector at those energies. If there is still an ambiguity, then the half-life of the radionuclide may provide a clue. For example, a radionuclide with a half-life of 1 min will not be present after a 1 h decay period. The final check on the origin of the gamma ray is a half-life measurement to confirm the identification of the source.

A computer based analyzer with data processing capability will most probably have a nuclide library for the identification of gamma ray energies. The program will check the energy of the unknown line against those of gamma rays falling within a preset tolerance of 1–2 keV. Not only will the system provide a list of possible sources based on gamma ray energy but it may also include checks on the presence of other lines from the same source and on the half-life of the source. Most commercial systems for gamma ray analysis will provide information in the form of a list of possible sources, followed by those rejected on the grounds of intensities and those rejected because the sample has been decaying for more than a preset number of half-lives. A final list of possible radionuclides in the sample is then supplied. In some cases the likely sources of gamma ray lines will be known, particularly if they are activation products. In that case the nuclide library could be limited to only those radionuclides that are going to be present in the sample.

It is important to remember that not all the lines in a gamma ray spectrum are the result of transitions during radioactive decay. Some lines occur as the result of secondary interactions in the detector such as the single and double escape peaks, annihilation peaks and summing peaks which will give lines in the spectrum which may be wrongly assigned. Sometimes these interfering lines are important sources of interference, for example the single escape peak from the 1778 keV line of ^{28}Al at 1267 keV could be mistaken for the gamma ray line of ^{31}Si at 1266 keV, since these single and double escape peaks are not usually listed in gamma ray identification tables. There are also naturally occurring gamma rays in the background, such as ^{40}K, which may be in the spectrum but not in the list. Nuclides with very short half-lives are sometimes omitted from compilations of gamma rays.

So the gamma ray energy is used to identify the radionuclide and hence the source of radioactivity. This qualitative gamma ray analysis may be sufficient for certain applications, where it is adequate to ascertain only what is present. Such examples would be for the identification of material or a source of pollution. Usually, however, once the radioactivity has been identified the next step is to measure it.

PEAK EVALUATION

The area under a peak in a gamma ray spectrum represents the number of counts collected for that gamma ray energy. The peak area divided by the count time will give the count rate for the gamma ray disintegration. Methods of evaluating the peak area vary but they all rely on subtraction of an estimated background value and only differ in the way that the assessment is made.

If the spectrum is simple and contains only single peaks, with no overlapping peaks or multiplets, then a simple background subtraction may be made. The simplest technique is where channels are selected on either side of the peak to represent an average value for the background, as shown in Figure 6.1. The contents of the two channels, c_a and c_i, are averaged and multiplied by the number of channels $(i-a+1)$, to give an estimation of the background to be subtracted from the total peak area:

$$\text{Peak area} = (c_a + c_b + c_c \ldots + c_i) - (c_a + c_i)(i-a+1)/2$$

The choice of background is entirely subjective and relies on there being little variation in the background counts. However if the peak area is large compared to the background and the background is flat, then it can give good results. A modification of the technique is to use smoothing to evaluate the background. For example five channels could be taken on each side of the peak and averaged to give a better estimate of the value to use. This will only be possible if there are five average channels available on either side. If the peaks are close together the second method may not be applicable. More accurate estimation of the background will be necessary if the spectrum does not present a relatively flat background either side of the peak.

Even the simplest computer based data processors will now include a

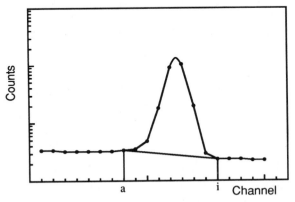

Figure 6.1. Evaluation of peak area using simple background subtraction. The total area under the peak is evaluated by summing the counts in channels c_a to c_i. The background is estimated by averaging the counts in channels c_a and c_i and multiplying by the number of channels $(i - a + 1)$.

peak search program of some description which will locate and evaluate peaks in the spectrum. The application of computer techniques to gamma ray spectrometry is reviewed by Op De Beeck and Hoste (1976), Yule (1981), and Debertin and Helmer (1988), including peak detection and evaluation through to quantitative analysis. Basically there are a handful of programs used in commercial systems for gamma ray analysis including HYPERMET, SAMPO, and GAMANAL (Phillips and Marlow, 1976; Koskelo et al., 1981; Gunnink and Niday, 1972). These computer programs use the Gaussian shape of the gamma ray peaks to fit the peaks, using exponential tails and a step function to cater for the change in height of the background from the low energy to high energy side of the peak. Computer programs are particularly useful for the separation of multiplets, using deconvolution techniques. Recently the programs have been modified for use on microcomputers and provide a powerful tool for routine analysis. Comparisons have been made on these programs and in most applications they all operate reasonably (Christensen and Heydorn, 1987).

Peak search reports generated by the computer programs have similar formats, regardless of the analyzer system used. The information provided will include the channel number of the peak centroid and the corresponding energy, the full width of the peak at half maximum height, the estimated peak area and the background that has been subtracted. The count rate is presented in counts per second, calculated by dividing the peak area by the recorded live time. The error on the estimated count rate is usually quoted as a percentage. It is calculated from the statistical variation (σ_S) in the estimated total counts in the signal peak (S) which is derived from the errors on both the total peak area which includes the background, ($S + B$), and the background (B):

$$\sigma_S = (S + B + B)^{1/2}$$
$$= (S + 2B)^{1/2}$$
$$\%\sigma_{rel} = 100 \, (S + 2B)^{1/2}/S$$

The limit of detection for a peak on a background is dependent on whether the background is well known or not. The limit of detection can be defined in a number of ways. It usually refers to the probability of 95% of a peak being detected above the background, and a value of about three times the standard deviation of the background is normally used. The background in multielement analysis is generally unknown and the value of 2.3 $B^{1/2}$, derived from the work of Currie (1968), is commonly used. However the limit of detection can be set on peak search programs by the operator and so the limit can be changed depending on the requirement. The limit of determination, the point at which a peak might be used for quantitative evaluation, is a higher figure defined by the standard deviation on the peak. A recent publication by Currie and Parr (1988) discusses practical detection limits in current use.

EFFICIENCY CORRECTION

If the count rate measured in a gamma ray spectrum is to be used to evaluate the activity of the source then the efficiency of the detector is required. The efficiency of a germanium detector, given in its specifications, will be quoted for the 1.33 MeV line of a ^{60}Co source counted at 250 mm from the detector, relative to that of a 76 mm × 76 mm sodium iodide detector. Consequently if an absolute measurement of the activity of a sample is to be made, it may be necessary to determine the absolute efficiency for the counting geometry of the sample at the range of energies that are in use. The resulting plot of detector efficiency against energy is the efficiency curve. It is important to note that in activation spectrometry it is not always necessary to determine the efficiency curve, for example if a sample and a standard are counted with identical geometry, the efficiency will be the same for both and they can be compared directly.

An efficiency curve is measured using sources of known activity or count rate. Reference sources are available from commercial suppliers, and they are calibrated to give the activity very accurately. Reference sources that are often used for the calibration are : ^{241}Am (60 keV), ^{133}Ba (81, 302, 356, 383 keV), ^{137}Cs (661 keV), ^{60}Co (1173, 1332 keV), ^{22}Na (1275 keV) and ^{88}Y (898, 1836 keV). For low energy gamma and X-ray energies, ^{57}Co (14.4, 122.1, 136.5 keV) is also useful. These references are normally available as solid point sources with a nominal activity of 37 or 370 kBq. Knowing the activity of the source on the day of certification, the activity of the source can be determined with the required detector geometry and the true activity can be calculated for the day on which it was measured.

For example, suppose a ^{133}Ba source with a certified activity of 370 kBq on 1 May 1987 is used to measure the detector efficiency at a fixed geometry on 1 May 1988. Knowing that the source has a certified activity of 370 kBq on 1 May 1987 and that the half-life of ^{133}Ba is 10.54 years, the activity on 1 May 1988 can be calculated:

Activity on 1 May 1987 = 370 kBq

$$\text{Activity on 1 May 1988} = 370 \times \exp(-\lambda t)$$
$$= 370 \times \exp(-t\ln2/T^{1/2})$$
$$= 370 \times \exp(-0.693 \times 1/10.54)$$
$$= 346 \text{ kBq}$$

The activity in disintegrations per second is then corrected for the branching ratio or gammas per disintegration to give the expected count rate for a particular gamma ray energy. For example, the gammas per disintegration for the 81 keV gamma ray is 34.3%. If the number of disintegrations per second is 346×10^3, to calculate the gamma ray emission rate for the 81 keV line:

$$\text{Gammas per second} = 0.343 \times 346 \times 10^3$$
$$= 1.2 \times 10^5$$

So the calculated gamma ray emission rate for the 81 keV line for the source on 1 May 1988 is 1.2×10^5.

The efficiency of the detector is measured by counting the source to determine the counts per second at 81 keV. Suppose the count rate is 120 counts per second:

$$\text{Efficiency} = \text{measured counts/emission rate}$$
$$= 120/1.2 \times 10^5$$
$$= 10^{-3}$$
$$= 0.1\%$$

An efficiency curve, like the one in Figure 6.2, can be plotted by making measurements at a number of energies distributed through the range from say 0.06–2 MeV. Computer based analyzers will have programs to do the calculations. The resulting efficiency curve can be used to calculate the detector efficiency for any gamma ray energy by interpolation. The efficiency curve only applies to one fixed geometry and the efficiency at a particular energy will vary depending on the detector distance. Figure 6.3 gives the series of plots for different detector distances.

It is therefore possible to determine the activity of a sample accurately provided that the efficiency of the detector is well known, which requires accurate determination of the efficiency curve using standard reference sources. The efficiency curve is unique to a particular detector and the

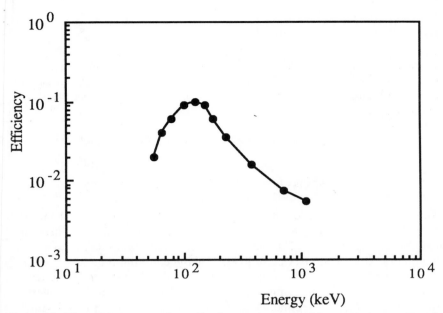

Figure 6.2. The efficiency curve for a Ge detector is produced by measuring sources of known activity at a range of energies. Computer based analyzers will calculate and plot the efficiency curve using the manufacturer's programs.

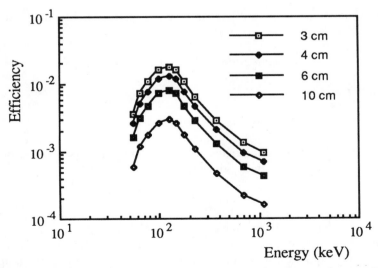

Figure 6.3. The detector efficiency curve is dependent on the source to detector distance. These curves relate to sources counted at 30, 40, 60 and 100 mm from the detector.

counting geometry and the quality of the data relies on the analyst. The count rate measured on the detector is then corrected for detector efficiency using the efficiency curve. The count rate then becomes true gammas per second

$$\text{Gammas per second} = \text{count rate} / E$$

QUANTITATIVE DETERMINATION OF ACTIVITY

The activity of a sample can be measured by gamma ray spectrometry, provided that the efficiency of the detector is known for the particular counting position used. The activity is calculated from peak area of the gamma ray line. The peak area divided by the counting time gives activity in counts per second, which must be corrected for detector efficiency at that energy to give the gammas per second. These have to be converted to disintegrations per second using the branching ratio (P) for the gamma ray of interest.

$$\text{Gammas per second} = (\text{counts per second}) / E$$

$$\text{Disintegrations per second} = (\text{gammas per second}) / P$$

Finally, the activity at time 0 (A_0) in disintegrations per second must be corrected for decay time (t_d) prior to counting:

$$A_0 = (\text{disintegrations per second})/\exp(-\lambda t_d)$$

To summarize,

$$A_0 = (\text{counts per second}) / E\, P \exp(-\lambda t_d)$$

If the half-life of the radionuclide is short it may also be necessary to correct by the factor for decay during counting time (t_c):

$$\lambda t_c/(1 - \exp(-\lambda t_c))$$

Any active source can be measured in this way on a gamma ray spectrometer. Environmental counting to monitor radioactivity in the environment requires only a further correction for the weight of sample to provide the Bq kg^{-1} in an active sample. For example, suppose that a 100 g source containing ^{137}Cs is counted by measuring the gamma ray at 661 keV on a detector in a geometry which gives an efficiency of 0.1%

at 661 keV. If the sample gives 3 counts per second this represents 3000 gammas per second. The branching ratio for ^{137}Cs at 661 keV is 85.0%, so 3000 gammas per second converts to 3000/0.85 disintegrations per second or 3.5 kBq. Dividing by the weight of the sample, the activity will be expressed as 35 kBq kg^{-1}.

In the case of neutron activation analysis the original sample is inactive and only becomes activated by irradiation in a neutron source. The activity is measured to deduce the amount of the element in the sample, using the activation equation.

"ABSOLUTE" ACTIVATION ANALYSIS

The activation equation derived in Chapter 2 can be solved to give the mass of the element using the measured count rate, knowing the value of the other factors:

$$w = A_0 \ A \ / \ N_A \ \theta \ \sigma \ \phi \ (1 - \exp(-\lambda t))$$

and since:

$$A_0 = (\text{counts per second}) \ / \ E \ P \exp(-\lambda t_d)$$
$$w = (\text{counts per second}) \ A \ / \ N_A \ \theta \ \sigma \ \phi \ E \ P \ (1 - \exp(-\lambda t)) \ \exp(-\lambda t_d)$$

The atomic weight, Avogadro's number and isotopic abundances are all well-known constants. The cross section, on the other hand is evaluated using measurements of a known mass of an element and the activation equation above. Uncertainties can be quite high particularly for some radionuclides with short half-lives. The decay constants and the branching ratios are usually known precisely but they may also be less precise in the case of short-lived radionuclides. Determination of the neutron flux and detector efficiency terms can only be made locally.

The value of the neutron flux is often the factor in the equation that is least well known, since it varies not only from source to source but also within the source itself. Precise measurement of the neutron flux is quite difficult and requires the use of monitors to measure it accurately. The flux is calculated from the activity measured in an irradiated foil using the activation equation. Care must also be taken so that the foil provides adequate information concerning the epithermal and fast components of the neutron flux. Detector efficiency curves must be well defined for the evaluation and care taken that the geometry is exactly the same in the reference sources and the flux monitor used, for accurate results.

Finally, it is usually possible to measure the length of irradiation very accurately, provided it is continuous. If the reactor has been shut down during the course of the irradiation, which may happen in the case of a reactor run on a daily basis, then corrections must be applied for decays between irradiation periods.

SINGLE COMPARATOR METHODS

Although it is possible in theory to calculate the mass of the element from a measurement of the activity, it is not usual practise to do so. Evaluation of the neutron flux every time a sample is irradiated would be cumbersome and it is more appropriate to monitor the flux by simply measuring the activity of a foil, in counts per second, and to compare it to the expected figure for a known flux to obtain a correction factor. It is not necessary to calculate the neutron flux value at all if the same value is to be used in the activation equation for an element of interest. The basis of the comparator method is that if the activation equation for the foil is compared to that of the element of interest, the flux values cancel out, together with the physical constant Avogadro's number and the following equation is obtained, where * refers to the comparator:

$$\frac{w}{w^*} = \frac{(\text{counts per second}) \, A \, / \, \theta \, \sigma \, E \, P \, (1 - \exp(-\lambda t)) \exp(-\lambda t_d)}{(\text{counts per second})^* \, A^* \, / \, \theta^* \, \sigma^* \, E^* \, P^* \, (1 - \exp(-\lambda^* t)) \exp(-\lambda^* t_d)}$$

The specific activity for an activation product is defined as:

$$A_{sp} = (\text{counts per second}) \, / \, w \, (1 - \exp(-\lambda t)) \exp(-\lambda t_d)$$

The ratio of the specific activities for the element of interest and the foil then becomes a simple expression which is called the k value (Girardi et al., 1965):

$$\frac{A_{sp}}{A_{sp}^*} = \frac{A^* \, \theta \, \sigma \, E \, P}{A \, \theta^* \, \sigma^* \, E^* \, P^*}$$

$$k = A_{sp} / A_{sp}^*$$

The specific activity for the element of interest can be calculated by multiplying the specific activity of the comparator by the k value. The k

value is only valid for a fixed sample detector distance so that E/E^* is constant from measurement to measurement. Also the effective cross section depends on the thermal to epithermal neutron ratio in the neutron flux and k is only valid on a single irradiation site. If different irradiation sites are to be used account must be taken of the epithermal flux component and the ϕ_{th} and ϕ_{epi} ratios considered (De Corte, 1969):

$$A_{sp} = \theta\, E\, P\, (\sigma_{th}\, \phi_{th} + I\, \phi_{epi})\, /\, A$$

$$k = \frac{A^*\, \theta\, E\, P\, (\sigma_{th}\, \phi_{th} + I\, \phi_{epi})}{A\, \theta^*\, E^*\, P^*\, (\sigma^*_{th}\phi^*_{th} + I^*\phi^*_{epi})}$$

A more general nuclear data value has been adopted which is independent of the reactor spectrum and of the characteristics of the detector. These new generalized factors are called k_0 values and may be considered as compound nuclear data (Simonits et al., 1975):

$$k_0 = \frac{A^*\, \theta\, P\, \sigma_{th}}{A\, \theta^*\, P^*\, \sigma^*_{th}}$$

In this case the atomic weight, isotopic abundance and cross section values are combined so that the ratio of the specific activity for the element of interest to the specific activity for the foil is a factor k_0. These factors will be constant regardless of the irradiation site. To make analytical use of the factor it is necessary to introduce the factors for the relative efficiencies of the lines used on the particular detector and to introduce the thermal and epithermal neutron flux components.

CHEMICAL STANDARDS

In most applications it is preferable to use a standard which contains the element which is to be determined. The reason is immediately obvious when the activation equation is considered. Although the activation expression appears to contain a number of factors for evaluation in fact it can be simplified if a standard is run in parallel with the sample. For example, if a standard containing a known mass of the element of interest was irradiated and counted under identical conditions as the unknown, all the factors would cancel out, leaving just the activity in counts per second:

$$w_{(SA)} = \text{counts per second}_{(SA)}\, A\, /\, N_A\, \theta\, \sigma\, \phi\, E\, P\, (1-\exp(-\lambda t))\, \exp(-\lambda t_d)$$

$$w_{(ST)} = \text{counts per second}_{(ST)}\, A\, /\, N_A\, \theta\, \sigma\, \phi\, E\, P\, (1 - \exp(-\lambda t))\, \exp(-\lambda t_d)$$

$$\frac{w_{(SA)}}{w_{(ST)}} = \frac{\text{counts per second}_{(SA)}}{\text{counts per second}_{(ST)}}$$

This simple relationship is only valid if all the other factors are held constant and sometimes it is not easy to ensure that they are. Obviously the cross sections, decay constants, atomic weights and isotopic abundances will all be identical. Even if the neutron flux is constant, the physical form of the standard must be the same as the unknown for the same flux to be seen by both the sample and standard. Irradiation, decay and counting times must be constant too and the counting geometry should be identical if they are to have the same counting efficiency. It is more usual for some of the conditions to be the same but for others to be different and for corrections to be made to the appropriate factors.

When chemical standards are used for comparison with unknown samples there are very few corrections to be made. If the activation conditions are the same and they are counted on the same detector, then usually the only corrections to be made are for different decay periods prior to counting and a flux correction due to irradiation position. The flux correction can be a simple procedure using single-element wires or foils, such as copper, iron, zinc or gold. Therefore where the samples and standards are irradiated in the same experiment, corrections are made for flux variation and decay time, but all other conditions remain constant.

Real samples, with known concentrations of the elements of interest, may be used as an alternative to chemical standards. They are favored by non-chemists who have not got the experience for preparation of accurate standards. In-house standards or bought-in reference materials may be used. Usually they are in nature as similar to the samples as possible, and well analyzed for the elements of interest.

If the measurements are repetitive in nature and the neutron flux can be monitored easily using a flux monitor, then it may not be necessary actually to run the elemental standards with every set of samples. Some analysts develop databanks for the activity induced in a standard for a particular length of irradiation. The activation data for the standards can be stored and recalled when required for comparison with the unknown samples. Changes in flux are corrected with the flux monitor data. Corrections can be made for irradiation, decay and counting time and detector efficiency curves may be used to correct for different counting geometry.

REFERENCES

Christensen, L. H. and K. Heydorn (1987), "Quality assurance in the determination of overlapping peak area," *J. Radioanal. Nucl. Chem.*, **113**(1), 19–34.

Currie, L. A. (1968), "Limits for qualitative detection and quantitative determination. Application to radiochemistry," *Anal. Chem.*, **40**(3), 586–593.

Currie, L. A. and R. M. Parr (1988), "Perspectives on detection limits" in L. A. Currie (ed.), *Detection in Analytical Chemistry. Importance, Theory and Practice.* ACS Symposium Series: 361, American Chemical Society, Washington DC.

Debertin, K. and R. G. Helmer (1988), *Gamma- and X-ray Spectrometry with Semiconductor Detectors,* North Holland, Amsterdam.

De Corte, F., A. Speecke, and J. Hoste (1969), "Reactor neutron activation analysis by a triple comparator method," *J. Radioanal. Chem.*, **3**, 205–215.

Erdtmann, G., and W. Soyka (1975), "The gamma-ray lines of radionuclides ordered by atomic and mass number. Part I, $z=2$–57 (helium–lanthanum),"*J. Radioanal. Chem.*, **26**, 375–495. "Part II, $z=58$–100 (cerium–fermium)," *J. Radioanal. Chem.*, **27**, 137–286.

Girardi, F., G. Guzzi, and J. Pauly (1965), "Reactor neutron activation analysis by the single comparator method," *Anal. Chem.*, **37**(9), 1085–1092.

Gunnink, R. J. and B. Niday (1972). *Report UCRL-51061. Vol I–IV.* Lawrence Livermore National Laboratory, Livermore, USA.

Koskelo, M. J., P. A. Aarnio, and J. T. Routti (1981), "SAMPO80: An accurate gamma spectrum analysis method for minicomputers," *Nucl. Instrum. Methods*, **190**, 89–99.

Op De Beeck, J. P. and J. Hoste (1976), "The application of computer techniques to instrumental neutron activation analysis," in T. S. West (ed.), *Physical Chemistry Series Two*, Butterworths, London.

Phillips, G. W. and K. W. Marlow (1976), "Automatic analysis of gamma-ray spectra from germanium detectors," *Nucl. Instrum. Methods*, **137**, 525–536.

Simonits, A., F. De Corte, and J. Hoste (1975), "Single comparator methods in reactor neutron activation analysis," *J. Radioanal. Chem.*, **24**, 31–46.

Yule, H. P. (1981), "Manual and computerized data processing in activation analysis," in S. Amiel (ed.), *Nondestructive Activation Analysis*, Studies in Analytical Chemistry 3, Elsevier, Amsterdam, pp. 113–137.

CHAPTER

7

SAMPLE PREPARATION

The quality of analysis using any technique is strongly dependent on the amount of care taken during the preparation of the sample. It is important that the person collecting and preparing the material for analysis is aware of the requirements of the technique, including the size and form of sample that can be accommodated. Impurities could be introduced into the material during preparation which would affect the analysis. Unless the sample provided is representative of the material to be examined the results may be meaningless. The requirement for representative analysis may mean that a large sample will be homogenized before subsampling and then some form of preconcentration may be necessary before the elements of interest may be determined.

SAMPLE SIZE

In the past it has been the sensitivity of neutron activation analysis that has been its main advantage. Consequently there has been a strong emphasis on the small size of sample required to determine trace elements. It is quite common to analyze samples of a few milligrams and it is of immense advantage in the analysis of medical samples or precious minerals, for example. Because of this emphasis on the need to analyze small amounts of material, many reactor irradiation sites have been developed to take only samples of small size and this has an effect on what may be analyzed.

In theory the amount of sample required is dictated by the sensitivity of the method for a particular element but in practice the background effects will be the main limitation. The factors affecting the induced activity for a particular target are the neutron flux, the mass of the element and the efficiency of the detector. In practice this means that the weight of sample, the length of irradiation and distance from detector may be varied to limit the radioactivity produced. Hence if one wishes to irradiate a large sample in order to get a representative analysis then the irradiation time can be shortened or the sample counted further from the detector to keep the count rate to a reasonable level. There are in fact several problems associated with analyzing samples weighing more than about one gram: the amount of activity produced may present a radiation hazard; neutron self-shielding may reduce the activity in matrices with high neutron cross sections; gamma ray attenuation may reduce the apparent activity in the sample; there may be difficulty in reproducing the geometry of a large sample in a standard. Consequently the sample size is usually kept below one gram.

By the same token, for a very small sample, such as a precious mineral grain or flake of paint from a valuable painting it is possible to extend the irradiation time or count for a very long time close to a very efficient detector to improve the sensitivity. If a short-lived radionuclide is being measured, then the irradiation time can only be extended to the maximum required to achieve saturation activity and the sample counted until all the activity has decayed. There are other techniques which can be used to counter this problem. For example, if short-lived radionuclides are being measured, cumulative neutron activation may be used to improve the analysis of a heterogeneous sample and cyclic activation may be used to improve the sensitivity for a small precious sample.

There is no upper limit on the amount of sample that may be activated and counted provided that the activity can be handled and there are occasions when one might wish to analyze a large sample. An example is in the determination of trace elements in pure polyethylene or carbon, where the matrix does not give activation products and the trace elements are in concentrations below $1 \mu g \, kg^{-1}$. The material is light and consequently a sample weighing only a few grams is quite bulky. Another example is the analysis of ore material as in the case of uranium or gold, where the elements are distributed heterogeneously. There may be advantages in analyzing up to 100 g of gold ore for representative analysis. If a 100 g sample is used it will be irradiated in a low neutron flux or for a very short time to control the activity of the sample.

In conclusion, there are a large number of constraints on the size of sample to be analyzed and the decision does not always rest with the amount of material available or the desired sample size. However in general it is possible to analyze a wide range of sample sizes, provided that the capacity of the irradiation facility is adequate, and special techniques can be applied to overcome the problems which occur as a result of the decision.

SAMPLE GEOMETRY

The geometry of the sample is a matter of choice, in the same way that it is for sample size, provided that it can be introduced into the irradiation site. Regardless of the form of the sample, whether it is solid or liquid, a reproducible geometry is most important. If the analysis is purely qualitative then a chunk of rock, a strand of hair or a flake of paint can be measured but if there is to be a comparison between samples and standards the geometry has to be very reproducible between them all.

The neutron flux distribution in the reactor irradiation site can be quite large and so there are advantages in keeping the samples and standards or monitors as close together as possible. The sample should not have too long a profile in the flux, otherwise there may be a significant neutron flux gradient across the sample itself. The flux distribution will also affect the activation if the shape is variable and the neutron absorption in the matrix will depend on sample thickness. In other words a short fat sample and a tall thin sample of the same material will not give the same activity on irradiation due to flux gradient and neutron absorption effects.

When it comes to counting, the differences will be even more significant as the efficiency of the detector falls off with distance. The smaller the sample, the closer it is to the detector when it is counted and therefore the sensitivity is greater. A tall thin sample will give fewer counts than an equivalent short fat sample. This can be demonstrated in a simple way in Figure 7.1 showing the efficiency of the counting geometry for a 5 cm³ sample of ion exchange resin on a 40 cm³ detector, using different pot diameters.

The amount of care necessary to get the most reproducible geometry will depend on the degree of accuracy required for the measurement. For example, if neutron activation is being used to analyze larvae to see if they have been feeding on a plant labelled with stable dysprosium, it may be adequate to irradiate the individuals without pretreatment to measure whether dysprosium is present in the larva. Even if the experiment is a quantitative one, the errors due to the different geometries of the larvae would probably be about 20%, which is well below other errors due to sampling. In mineral exploration it may be possible to analyze drill chips in bulk and accept errors of say 20%, in order to speed up the process in the initial stages of gold exploration.

For quantitative analysis care must be taken to present the sample in a reproducible form. Rocks and ores are usually crushed and homogenized for sampling anyway, but mineral grains resulting from separation

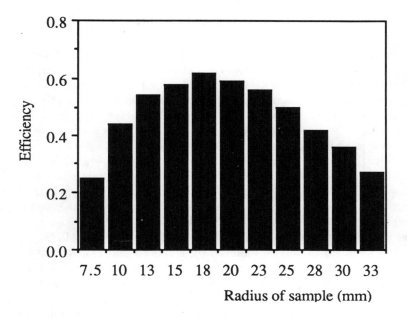

Figure 7.1. The relative counting efficiency for a 0.005 dm³ sample is dependent on the radius of the container. This plot shows that a container with a radius of 18 mm has the most efficient geometry.

procedures can be left alone, provided they are reasonably small. Medical samples are usually dried and ground, and so is plant material. Liquid samples should be taken as a fixed volume. Air filter paper samples should be folded in a fixed geometry, and so should the filter papers from preseparation procedures. Most of these preparative stages are part of the sampling procedure and consequently samples are fairly homogeneous when ready for analysis. Reduction of size may be required because of limitations in the irradiation site, particularly for plant material and some medical samples which can be quite bulky. Small samples are more convenient and it is possible to compact plant, medical and filter paper samples by pelletizing or ashing and pelletizing. Pelletized samples can present an extremely reproducible geometry.

In conclusion, the geometry of the sample is dictated by several factors which may not be overcome so readily as the size problem, but as a general rule the geometry must be chosen so that it is reproducible and so that a suitable standard may be presented in the same form.

SAMPLE COLLECTION

Once the limitations on size and geometry have been established the mode of sampling can be defined. Most of the problems of sampling are common to all analytical techniques used for the determination of trace elements. In any case samples are routinely collected for analysis by more than one method so it is rarely a question of seeking the optimum technique. The main problems are obtaining a representative sample and avoiding external contamination on collection.

Heterogeneity is a problem common to all fields of study and it is important to obtain a representative sample for analysis whether it is a lump of rock, a piece of human liver or a sack of polyethylene beads. The size of sample initially required and the size of the subsample necessary for analysis are defined by the sampling constant, a term which was defined for geochemical use by Ingamells and Switzer (1973) and has been applied to biological material by Heydorn and Damsgaard (1987). It is often possible to take large samples for rock analysis and quite usual to crush and grind kilograms of material for analysis, regardless of the analytical technique. Taking a small sample from the bulk material must be done carefully and the techniques of sampling rock material are described well by Potts (1987). Even after careful preparation it is possible to see heterogeneities in material where distribution of the elements is not uniform.

In the area of environmental samples it is also important to collect large samples for a representative analysis. This is feasible in the case of vegetation such as leaves, grasses and mull samples. Air and river water may be sampled over long periods of time to provide a sufficiently representative sample, provided there are no frequency variations. In clinical research the homogeneity of the sample is dependent on the matrix and although fluids such as blood are quite uniform, organs such as liver and kidney are heterogeneous. The specific problems of sampling in clinical research are addressed by Heydorn (1984) who indicates, for example, that a 200 g sample of liver or kidney is required to reduce to 1% the variation in the results for replicates.

Contamination during sampling may be more important in activation analysis than other techniques because there is less handling of the sample before analysis. Geological samples are normally obtained by drill-core sampling at an exploration site or on a smaller scale by a geologist using a hammer. Contamination from the metal of the tool can occur and it is wise to clean the outside of the sample with a diamond buffer and wash with deionized water prior to crushing. It is not a good idea to use acid to wash the sample in case it preferentially leaches out any elements. Metal contamination is the main problem whatever the sample type. If plant material is harvested with metal agricultural tools there will be a danger of pickup, and medical samples are normally obtained using stainless steel scalpel blades or hypodermic syringes which can result in contamination from iron, chromium and cobalt in particular. In many cases there are no suitable alternative tools available but in the medical field, where the trace elements to be determined are present in extremely low levels, it is preferable to cut tissue with a quartz knife (Lux et al., 1987) and to collect blood with a plastic catheter. The special problems of collecting blood samples have been described in detail elsewhere (IAEA, 1984; Versieck, 1984; Speecke et al., 1976; Ibbott, 1976). Plastic tweezers should be used to handle delicate samples. The specific problems of sample collection are discussed fully in several publications on the general topic of contamination of clinical samples (Cornelis, 1987; Heydorn, 1984; Iyengar, 1984; Sansoni and Iyengar, 1978; Zeif and Speights, 1972; Zief and Mitchell, 1976) including two special publications on the subject of sampling and analysis of biomedical samples (LaFleur, 1976; IAEA, 1984).

The container used to collect the samples can be a source of contamination and much use is made of plastics in general and polyethylene in particular. They have very low concentrations of trace elements and therefore they are not a major source of contamination themselves. Another reason for using plastics is that they are disposable and therefore new, unused containers can be used each time. Liquid samples in particular are likely to interact with the container and losses can occur onto the surface of the bottle, particularly with glass walls. Storage of the sample in the container can result in losses and contamination. Long-term storage is a particular problem in the case of water samples (Qian et al., 1987). A study of the storage of medical samples in polyethylene containers showed that the effects are significant (Heydorn and Damsgaard, 1982). Some elements such as mercury have been shown actually to diffuse through the walls of polyethylene containers on standing. In addition to losses and contamination problems it is also possible that material of an unstable nature may undergo a chemical change on standing and losses may occur as a result of that.

It is clear that the analyst must ensure there is no danger of the sample being contaminated with an element of interest during the collection and preparation of the sample. Therefore it is essential to control the mode of sampling and to make known to the analyst any details so that steps may be taken to clean the sample if contamination appears to be possible.

HOMOGENIZATION

Homogenization may be considered to serve two purposes. The first is to provide a uniform source of material from which a small aliquot may be taken that will be representative of the material. Clearly, if a sample is to be analyzed it must be representative of the material it comes from, whatever analytical method is used. However, in activation analysis the size of the sample taken may be very small and it is important to ensure that it is representative.

The second reason for homogenizing a sample is to ensure that the actual sample to be irradiated and counted is uniform throughout. In other comparable analytical techniques for trace element analysis the final sample is usually in liquid form and so by nature is uniform. Taking a sample of a liquid for activation analysis is not a problem but if, as is more usual, a solid is used it must be homogeneous. Suppose, for example, a rock sample contains a few grains of a platinum mineral. The activation of the platinum will depend on the position of the mineral grains in relation to the neutron flux. The count rate from the activated platinum mineral will depend on the distance of the grain above the detector and whether it is at the top or bottom of the sample. It is much more satisfactory if the mineral grains are ground finely and dispersed uniformly throughout the sample.

Homogenization is achieved by processing all the material collected. In general solid samples are ground to a fine powder. Coarse crushing is used for the preliminary reduction of rock or archaeological material, followed by grinding to a small mesh size. These methods of preparation may add to the sources of contamination if care is not taken during the process and the problems are discussed in detail by Flanagan (1986) in a report on the preparation of geochemical reference samples. Rock or drill-core samples are crushed roughly to reduce the size sufficiently for grinding. Steel jaw crushers do not appear to present a problem and are preferable to the use of tungsten carbide tools. Contamination from tungsten not only precludes that element from the analysis but the neutron activation of tungsten produces ^{184}W which has a half-life of 24 h and can cause large interfering peaks and background activity in the gamma ray spectrum. It is also possible that tungsten carbide grinders can cause tantalum contamination and tantalum is an element of major interest in mineralogical problems, at concentrations below mg kg^{-1}. For grinding rocks and archaeological clay materials it is advisable to use agate to avoid contamination, although occasionally agate can contain contaminating mineral veins. The grinding tools can be tested by putting blanks through them, such as pure sand, to check for contamination but it is difficult to reproduce the operation exactly and almost impossible in the case of the jaw crushers.

Plant material and biological tissue are also ground down to a fine powder after drying. The samples themselves have to be dried before homogenization using low temperatures to avoid loss of trace elements. Medical samples, including tissue and blood samples are more usually lyophilized (freeze dried); this process involves freezing the sample to about $-20°$ C before putting it under vacuum in the freeze-drier to remove the water by sublimation. Vegetation is dried using a low temperature oven, perhaps under vacuum or by using lyophilization. Grinders and blenders for vegetation and biological samples generally have steel blades which may cause contamination, particularly with hard samples. PTFE vessels and balls are now used for the homogenization of biological materials (Iyengar, 1976).

PRECONCENTRATION

A sample may be considered to be preconcentrated if part of it is removed prior to irradiation. The reduction in volume will automatically improve the sensitivity of counting on the detector due to the improved geometry. It may also be necessary to reduce the size so that it will fit into the irradiation site. For liquid samples such as blood, urine or environmental water it will generally mean the drying of a sample by removal of the water content. Lyophilization techniques are normally used for the removal of water, although evaporation by heating can be used if there is no loss of elements of interest during the process.

For solids the reduction in volume is more difficult and usually confined to the low temperature ashing of vegetation to drive off unwanted carbon, oxygen and nitrogen. The reduction in volume is substantial. Further reduction in volume can be applied to ashed samples using pelletizing but there is no increase in elemental concentration during the process. A pellet is much easier to handle than a fine powder both before and after irradiation.

Preconcentration techniques can also include removal of unwanted elements in the matrix of the sample. Such separation steps before irradiation lose the main benefit of activation analysis, which is that once the sample is activated any chemistry will not cause contamination of the sample. However there are situations where preconcentration is necessary, such as where the radionuclide is very short-lived and there would not be time to separate or where the post irradiation separation would be hazardous to the analyst. Ion exchange separation techniques are generally used in this type of work. Also, in the special case where identification of the species is required it is usual to carry out chemical separation of the species prior to analyses. Examples of such work include the analysis of selenium to elucidate the role of the element in the human body.

In the case of rock material, preconcentration usually means the digestion of the matrix followed by ion exchange or solvent extraction separation. For example, the rare earth elements are collected on an ion exchange column and then eluted in sequence. In the particular case of the platinum group elements and gold, fire assay preconcentration is applied to reduce the sample size and to collect the elements into a small button of lead or nickel sulfide. This technique of fire assay collection is an ancient method of concentration which is applied to precious metal analysis regardless of the final analytical method applied.

Preconcentration techniques, like the separation methods applied after irradiation, are not specific to activation analysis and consequently are better described in the relevant literature, such as the recent publication on ion exchange separation methods by Korkisch (1988).

REFERENCES

Cornelis, R. (1987), "Sample handling of clinical specimens for ultratrace element analysis," *J. Radioanal. Nucl. Chem.*, **112**(1), 141–150.

Flanagan, F. J. (1986), "Reference samples in geology and geochemistry," *US Geological Survey Bulletin 1582*, US Government Printing Office.

Heydorn, K. (1984), *Neutron Activation Analysis for Clinical Trace Element Research, Vols I and II*, CRC Press Inc.,Boca-Raton, Florida.

Heydorn, K. and E. Damsgaard (1982), "Gains or losses of ultratrace elements in polyethylene containers," *Talanta*, **29**, 1019–1024.

Heydorn, K. and E. Damsgaard (1987), "The determination of sampling constants by neutron activation analysis," *J. Radioanal. Nucl. Chem.*, **110**(2), 539–553.

IAEA (1984), *Quality Assurance in Biomedical Neutron Activation Analysis*, IAEA-TECDOC-323, International Atomic Energy Agency, Vienna.

Ibbott, F. A. (1976), "Sampling for clinical chemistry," in P. D. LaFleur (ed.), *Accuracy in Trace Analysis: Sampling, Sample Handling, Analysis. Volume I.* NBS Spec. Pub. 422, National Bureau of Standards, Washington, pp. 353–361.

Ingamells, C. O. and P. Switzer (1973), "A proposed sampling constant for use in geochemical analysis," *Talanta*, **20**, 547–568.

Iyengar, G. V. (1976), "Homogenised sampling of bone and other biological materials," *Radiochem. Radioanal Lett.*, **24**(1), 35–42.

Iyengar, G. V. (1984), "Preservation and preparation of biological materials for trace element analysis: quality assurance considerations," in IAEA, *Quality*

Assurance in Biomedical Neutron Activation Analysis, IAEA–TECDOC–323, International Atomic Energy Agency, Vienna, pp. 71–82.

Korkisch, J. (1988), *Ion Exchange Resins: Their Application to Inorganic Analytical Chemistry*, CRC Press, Boca Raton, Florida.

LaFleur, P. D. (ed.) (1976), *Accuracy in Trace Analysis: Sampling, Sample Handling, Analysis, Volume I*, NBS Spec. Pub. 422, National Bureau of Standards, Washington.

Lux, F., T. Bereznai, and S. Trebert Haeberlin (1987), "Minimization of the blank values in the neutron activation analysis of biological samples considering the whole procedure," *J. Radioanal. Nucl. Chem.*, 112(1), 161–168.

Potts, P. J. (1987), *A Handbook of Silicate Rock Analysis*, Blackie, Glasgow.

Qian, X. Z., X. X. Li, X. Y. Mao, and C. F. Chai (1987), "An investigation of concentration variations of trace elements during storage of natural water samples," *J. Radioanal. Nucl. Chem.*, 113(1), 77–82.

Sansoni, B. and V. Iyengar (1978), "Sampling and sample preparation methods for the analysis of trace elements in biological material." IAEA–SM 227/107, International Atomic Energy Agency, Vienna, 1978.

Speecke, A., J. Hoste, and J. Versieck (1976), "Sampling of biological materials," in P. D. LaFleur (ed.), *Accuracy in Trace Analysis: Sampling, Sample Handling, Analysis. Volume I.* NBS Spec. Pub. 422, National Bureau of Standards, Washington, pp. 299–310.

Versieck, J. (1984), "Collection and manipulation of samples for trace element analysis: quality assurance considerations," in IAEA, *Quality Assurance in Biomedical Neutron Activation Analysis*, IAEA–TECDOC–323, International Atomic Energy Agency, Vienna, pp. 83–106.

Zeif, M. and R. Speights (1972), *Ultrapurity, Methods and Techniques*, Marcel Dekker, New York.

Zeif, M. and J.W. Mitchell (1976), *Contamination Control in Trace Element Analysis.* Chemical Analysis Series, Volume 47, Wiley-Interscience, New York.

IRRADIATION CONTAINERS

The nature of the irradiation container can have a significant impact on the quality of the activation analysis process. The shape and size of the container and the type of irradiation to be carried out will dictate the material chosen. If the samples are counted in the irradiation container there may be impurities present in the material which affect the results. Totally automated analyses will be affected by pickup of impurities during irradiation. The material of the irradiation container, be it an absorber or a moderator, will change the neutron flux seen by the sample. The important role of the sample container is frequently ignored and although these effects may be small they do contribute to the quality of data obtained by activation analysis.

SHAPE AND SIZE

The shape and size of the irradiation capsule will depend on the irradiation site in which the sample is to be activated. Most irradiation facilities, whether they are associated with neutron, photon or charged particle sources, are built individually and they have a unique design. In general the diameter of the irradiation tube is often limited by what size is available in the quality of material used, such as high grade aluminum or stainless steel. There are a few commercial irradiation systems available which are designed for particular size containers but it is common to have more than one size irradiation tube on the same reactor. In accelerators the target holder is designed with similar criteria for materials. If metals or semiconductors are being irradiated with charged particles it may not even be necessary to encapsulate the sample at all.

There are a wide variety of container sizes used for irradiations but in general they are cylindrical in shape, the height and diameter being dictated by the irradiation site. In a pneumatic transfer system, where the capsule has to fly in the irradiation tube, the shape of the irradiation rabbit is often not cylindrical but wider in the center narrowing at the ends or a dumb-bell shape with a larger diameter at the ends of the rabbit, as shown in Figure 3.2.

It is usual for reactor operations staff to insist on double containment of the sample in case of leakage, so smaller capsules are designed to fit tightly inside the outer irradiation container. It is possible to accommodate more than one inner capsule in the outer container and quite usual to irradiate 10 or more samples at once.

There are other considerations which concern the analysts and which must be addressed despite the limitations on capsule design enforced by reactor operations considerations. When a liquid or a powdered solid sample is irradiated in a capsule the geometry of the sample will be that of the container. Consequently the container geometry must be reproducible, that is of constant wall thickness. If the container is very much larger than the sample there is a danger that the sample will be spread round the walls of the container. Therefore for the containment of small samples it is necessary to use sufficiently small inner containers to prevent the sample from moving about. Standards or flux monitors can be included with samples in the same irradiation container. These inner containers may be placed one above another to form a cylinder of samples inside the outer irradiation container. Alternatively the samples may be packed in long thin tubes which stack side by side to form a bunch of rods in the container. The appropriateness of the configuration will depend on the flux distribution in the irradiation container. In general it is probably easier to correct for flux variation in the cylindrical geometry than in the rod-like geometry.

Consideration must be given to how the capsule is to be sealed prior to irradiation and this may also affect the style of capsule used. It may be a requirement to seal the containers to be sure of proper closure. If the irradiated sample is to be transferred to a clean container for counting then some thought must be given to how easily the capsule can be opened. A screw thread is easy to undo, particularly if remote handling tools are used. A snap-on lid is also easily undone but will probably have to be sealed during irradiation to avoid leakages.

On the other hand it is quite usual to count the sample in the same container used for irradiation, provided that there will be no contamination problem from the material of the container. If the sample is to be counted in the same container then the shape must also suit the counting geometry requirements. A short cylindrical container will be more suitable for counting than a tall thin capsule, for example.

MATERIAL

The irradiation capsule should be made of a material which does not activate too much and which can withstand any radiation damage or heating effects which occur during irradiation. Apart from those special requirements the material of the container will have to be physically strong, easily manufactured into the capsule and not expensive. The materials which fit those requirements and are commonly used are aluminum, polyethylene and quartz glass.

Aluminum is used in the manufacture of reactor parts generally because it does not corrode and activates to produce only a short-lived radionuclide. It is not activated to any extent in charged particle bombardment. It is convenient in foil form to wrap samples and may be manufactured into irradiation cans. Aluminum containers are used for the irradiation of long-lived radionuclides which require activation with a high integrated neutron dose and do not need to be unloaded immediately after irradiation. Aluminum capsules do have to be manufactured for irradiation work and an example is the Harwell "A" can which has a screwtop lid with a slot for remote handling tools to remove it. The containers are not cheap because they are specially manufactured but they can be used many times without build-up of radioactivity.

It is common to use polyethylene for the manufacture of irradiation capsules because the material does not produce significant radioactivity on irradiation with reactor neutrons and the material is widely used in the manufacture of disposable containers. There are advantages in using a disposable material for radioactive work and the material is cheap. Polyethylene containers are produced commercially in many shapes and sizes and they can be purchased very cheaply. Polyethylene is easily molded into containers and there are commercial companies who are willing to manufacture capsules to the customers' specifications. The cost is not high once the mold has been purchased. Polyethylene is not very easily machined, if for example the lids have to screw on but low density polyethylene can be heat sealed using an ordinary soldering iron so that the container can be made leakproof. The main disadvantage of polyethylene is that it cannot withstand radiation damage at high neutron doses. After a while the polyethylene will go brown, it becomes quite brittle and is liable to break up. This means that polyethylene is used for short irradiations of up to one week in low power research reactors. Of course polyethylene cannot be used to encapsulate samples that are being analyzed for light elements such as carbon by photon activation and the material would not withstand the temperatures induced in charged particle analysis.

In some special cases, particularly where elements are likely to be lost through the walls of polyethylene containers during irradiation, as in the case of selenium and mercury (Heydorn and Damsgaard, 1982), quartz glass is used for sample irradiation. It is clearly a material which is only suitable for an inner container, enclosed in aluminum or polyethylene. Glass is often used where the sample is to undergo chemical separation after irradiation and the glass ampoule is crushed and digested with the sample. Impurities can occur in the quartz and on the surfaces and careful washing with acid is necessary. Sealing these glass ampoules can create problems and there is even a danger of contamination at that stage (Lux et al., 1987).

Graphite has been used to overcome the problem of thermal decomposition (Chisela et al., 1986). Graphite has the advantages of being light, durable, unaffected by temperature and obtainable in a form sufficiently pure to allow counting in the same container. A disadvantage is that the capsule is not easily sealed.

IMPURITIES

The purity of the irradiation capsule is affected by the impurities present in the raw material used to manufacture the capsule, the impurities introduced during manufacture, contamination on handling after manufacture and pickup during use. Some of these stages are beyond the control of the analyst but much can be done to minimize the effects of these impurities. The effects are going to be significant in the case of neutron activation and less important for photon and charged particle activation where the surface of the sample is usually etched or machined after irradiation and before counting.

If the sample is to be transferred to a clean container after irradiation there is no problem from impurities in the capsule, only from contamination on the surface of the container. Contamination on the surface of the capsule may have been introduced during the manufacture stage, particularly due to handling of the product which will contaminate it with sodium and chlorine. If the handler is wearing jewelry, elements such as gold and iridium can be transferred to the surface and any cutting and machining tools can contaminate with chromium and titanium for example. Some of these problems may be avoided by careful washing of the capsule before use and it may be necessary to use nitric acid, deionized water and acetone to clean surfaces properly. It is also advisable to wear plastic disposable gloves or use clean plastic tongs to handle the capsule after it has been washed. The effect of surface contamination will be most significant in the case of liquids but solids can be affected and this creates a problem which is of particular importance in clinical research, due to the low levels of trace elements to be determined. Heydorn and Damsgaard (1982) have investigated the problem of surface contamination in polyethylene containers using a variety of washing techniques including water, nitric acid and hydrogen peroxide. Manganese, which was present at about 20 ng in the container, was released from the walls of a polyethylene vial during irradiation. The amount was small, less than 1 ng, but variable so that no accurate estimation of the blank could be made. The surface contamination on quartz ampoules is only satisfactorily removed using hydrofluoric acid (Lux et al., 1987).

Since neutron activation is a routine method of analysis, capable of processing large numbers of samples, it is preferable that sample handling is minimized after irradiation. It is quite usual to leave the sample in its capsule for counting, after removing it from the outer irradiation container. If the sample is to be counted in its inner irradiation capsule then the impurities in the material of the capsule must be considered. The impurities may cause a blank value which results in erroneous results or may simply cause a high background and raise the detection limit of the determination. Aluminum will create a high background count immediately after irradiation and so it is not suitable for analysis of short-lived radionuclides. Other impurities in aluminum, such as manganese, will cause a background effect even after the aluminum has decayed. Quartz glass is sometimes digested along with the sample for radiochemical separation and then the impurities in the material are important. There are no materials used for irradiation capsules which are entirely free of impurities. Polyethylene is probably the only material that can be used in this way and it is not totally free from trace elements. Table 8.1 shows the impurities in encapsulation material, namely aluminum (Mizohata et al., 1978), polyethylene (Benzing, unpublished data) and quartz (Lux et al., 1987).

Low density polyethylene, in the form of the beads used for injection molding, is relatively pure. The trace elements which can cause problems are Al, As, Br, Cl, Cu, Fe, K, Mo, Na and Zn (Kucera and Soukal, 1983). The aluminum appears to be distributed uniformly but the other elements are variable and therefore cannot be corrected for. The impurities introduced during injection molding will depend on the other work going on in the workshop. Capsules containing nanograms of gold were found to result from manufacture in a factory where precious metals were machined (Parry, 1980).

If the activation analysis process is totally automated then the sample will be transferred directly from the irradiation site to the counting

Table 8.1. Concentrations of trace elements in high purity material used for irradiation containers (in $\mu g\ kg^{-1}$)

Element	5N class aluminum[a]	Low density polyethylene[b]	Suprasil quartz[c]
Na	320	<80	
Mg		<500	
Al		100	100
Cl		1900	
Ca		<700	<160
Sc	120	<0.1	
Ti		<200	
V		<0.3	
Cr	40	<41	0.44
Mn		<2	10–30
Fe	3000	<840	13
Co	2.5	<2	0.0065
Ni		<6000	0.39
Cu	3000	<40	0.1–2
Zn	<3000	<62	3.8
Ga	<40	<1.3	
Ge		<400	
As	17	<0.6	
Se	<30	<200	0.0046
Br	<10	22	70
Rb		<28	<0.032
Sr		<300	
Nb		<700	
Mo		<200	
Rh		<0.2	
Ag	<50	<20	<0.07
Cd	<0.1	<300	
In		<0.06	
Sn		<200	
Sb	50	<0.4	3.8
I		<6	
Cs	<6	<1.2	<0.002
Ba	<2000	<64	
La	140	<0.2	
Ce	250	<5.2	
Nd		<7.9	

Table 8.1. Continued

Element	5N class aluminum[a]	Low density polyethylene[b]	Suprasil quartz[c]
Sm	17	<0.4	
Eu		<0.02	
Gd		<5.7	
Tb		<0.4	
Dy		<0.5	
Ho		<0.2	
Yb	8	<0.4	
Lu	2.1	<0.07	
Hf	<6	<1.3	
Ta		<1.3	
W	10	4.8	
Au		<6	0.13
Th	69	<0.7	
U		<0.7	

Sources: [a] Mizohata et al., 1978
[b] Benzing, unpublished data
[c] Lux et al., 1987; Heydorn, 1984

position. If the sample is to be counted in the irradiation container, it must be free of impurities. A further problem in this case is the pickup of trace metals in the irradiation site. The aluminum used for the irradiation tube can contain a high level of manganese and both these elements can contaminate a polyethylene rabbit travelling round tight bends. The problem is overcome by ensuring that bends are made in a material which does not contain metal impurities, such as polyethylene, if it is outside the reactor core, or to use a complete carbon fiber transfer system (Bode and de Bruin, 1988). Pickup can also occur in the sample loading device if it is made of metal.

The presence of impurities can be checked using blank determinations, although it is sometimes difficult to pinpoint the actual source of the contamination. A systematic evaluation may be necessary starting with the raw material and working through to the irradiation site. If the impurity is present in a constant amount then it is probably in the raw material and it may be possible to correct the samples for that component. If the contamination is not consistent it may have been introduced during manufacture, or caused by handling. Pickup during irradiation will be confirmed if the level of impurity increases each time the capsule is irradiated. If pickup does occur it may be impossible to make an accurate determination of that element without transferring the sample to a clean container.

EFFECT ON NEUTRON FLUX

The effect of the irradiation capsule itself on the activation of the sample it contains is frequently overlooked. The material of the container is placed between the neutron source and the sample and is bound to have some effect on the neutron flux. Moderation of the neutron flux is enhanced by polyethylene with its high hydrogen content and so a polyethylene container with a thick wall will moderate the higher energy component of a reactor neutron spectrum to a greater extent than one with a thin wall. As a result the sample will experience a neutron flux with a higher thermal neutron component, which will actually enhance the activation of many elements of interest. It is therefore important that all samples and standards are encapsulated in exactly the same containers and that they are packed in a uniform way. In other words a batch of long thin samples placed vertically as a bundle in a cylindrical container on one side of a reactor core does not present a uniform thickness of encapsulation material between all the samples and the neutron flux; some samples will be irradiated under a greater thickness of moderator (polyethylene) or absorber (aluminum). If a flux monitor is included with the samples it should be enclosed in exactly the same type of container or the flux measured will not correspond to the flux seen by the sample. These effects may be quite small but should be considered.

If a suitable absorber is introduced deliberately then the thermal component of the reactor neutron flux may be removed completely. This is the case for thermal neutron filters of cadmium or boron. A cadmium sheet 1 mm thick will remove all the neutrons with energies below 0.5 eV and so if the sample is irradiated under a cadmium filter it will only be exposed to epithermal and fast neutrons. Boron will also remove thermal neutrons and can be used in the same way. Irradiation tubes incorporating the thermal neutron filters can be installed in the reactor but if such a facility is not available the same effect can be obtained by designing an irradiation container with a wall of absorber. Cadmium 1 mm thick is convenient because it takes up little space but the activation of the cadmium itself causes problems if the sample is to be unloaded soon after irradiation. The alternative is to use boron or boron carbide which present no problems of activation, but are more difficult to machine. The containers are described in Chapter 3 and shown in Figure 3.8.

REFERENCES

Bode, P. and M. de Bruin (1988), "An automated system for activation analysis with short half-life radionuclides using a carbon fiber irradiation facility," *J. Radioanal. Nucl. Chem.*, **123**(2), 365–375.

Chisela, F., D. Gawlik, and P. Bratter (1986), "Evaluation of high purity graphite sample containers for use with rapid instrumental epithermal neutron activation analysis," *J. Radioanal. Nucl. Chem.*, **102**(2), 347–357.

Heydorn, K. (1984), *Neutron Activation Analysis for Clinical Trace Element Research*, Volume I, CRC Press, Boca Raton, Florida.

Heydorn, K. and E. Damsgaard (1982), "Gains or losses of ultratrace elements in polyethylene containers," *Talanta*, **29**, 1019–1024.

Kucera, J. and L. Soukal (1983), "Determination of trace element levels in polyethylene by instrumental neutron activation analysis," *J. Radioanal. Nucl. Chem.*, **80**(1–2), 121–127.

Lux, F., T. Bereznai, and S. Trebert Haeberlin (1987), "Minimization of the blank values in the neutron activation analysis of biological samples considering the whole procedure," *J. Radioanal. Nucl. Chem.*, **112**(1), 161–168.

Mizohata, A., T. Mamuro, and T. Tsujimoto (1978), "Multielement trace analysis of high purity aluminium by neutron activation," *J. Radioanal. Nucl. Chem.*, **42**, 143–152.

Parry, S.J. (1980), *Evaluation of Low Density Polyethylene for Irradiation Capsules*, University of London Reactor Centre Internal Report, unpublished.

PREPARATION OF STANDARDS

Unknown samples are irradiated alongside some kind of standard for quantitative analysis, be it a chemical standard, a reference material or a flux monitor. Even if a flux monitor is used it is generally compared to a database formed by the irradiation of standards on a previous occasion. Consequently at some stage during the setting up procedure chemical standards are used and in many respects they provide the best method of standardization in quantitative analysis. Whichever type of standard is used in the evaluation, care must be taken to reproduce the geometry of the sample and ensure that the flux is also reproducible or can be corrected for.

SINGLE-ELEMENT STANDARDS

In its simplest form a chemical standard can consist of one element. The concentration should be of the same order of magnitude as the expected concentration in the sample but, unlike other techniques, because of the direct proportionality between activity and concentration it is not necessary to compile a calibration curve or reproduce concentrations exactly. The standard can be in the form of an element or a compound depending on its chemical and physical properties. Laboratory grade chemicals will contain substantial levels of other trace elements and care must be taken to ensure that these trace impurities do not interfere with the analysis. It may be necessary to use high purity chemicals. The choice of chemical form of the element will depend on the nature of the high purity material available and on the type of irradiation to be used. With chlorides, for example, the activation of ^{38}Cl could cause interferences on a short irradiation but would not affect analysis after a few hours decay. Sulfates and nitrates are the other most widely used soluble compounds, nitrogen, sulfur and oxygen do not activate but the use of nitrates may be restricted in the reactor due to their unstable nature. It must also be remembered that not all chemical compounds are stoichiometric and care should be taken to ensure that the figure used to calculate the proportion of the element in a compound is correct. There is also a possibility that the material is non-isotopic. Uranium compounds are usually depleted in ^{235}U when provided by chemical companies, which could have a significant effect on the determination of uranium by delayed neutron counting but would have little effect if gamma ray spectrometry is to be used, with the $^{238}U(n,\gamma)^{239}U$ reaction.

When neutron activation analysis is used to measure trace elements the standard will need to be in low concentrations for comparison with the sample. Apart from weighing out microgram quantities of the pure element or compound, which is bound to introduce errors, the standard can be diluted to bring it down to a sufficiently low concentration. For example, if a rock sample weighing 0.1 g is to be analyzed for cobalt and it is expected that the rock contains about 1 g kg^{-1} of cobalt, then a standard containing 0.1 mg of cobalt is required. It could be difficult to weigh out 0.1 mg of cobalt very accurately but a solution containing say 1 g dm^{-3} of cobalt would be readily made up by dissolving 1 g of cobalt, as a soluble compound, in 1 dm^3 of dilute acid. Alternatively there are standard solutions available commercially containing 1 g dm^{-3} of the elements made up for atomic absorption spectrometry or inductively coupled plasma emission spectrometry. Then 100 mm^3 of solution, containing 0.1 mg of cobalt can be accurately dispensed with a micropipette. Commonly the solution will be pipetted onto a filter paper and dried before use.

Care must be taken with the storage of dilute solutions since they may deteriorate in time. Processes that can affect the concentration of a solution on standing include adsorption of elements onto the wall of the container, precipitation of dissolved solids and chemical breakdown of species such as conversion of iodide to iodine. It is helpful to store the solution in acid but preferable to transfer the solution as soon as it is prepared into the final solid form required for analysis. Since the most accurate way of preparing a chemical standard is in liquid form and the most usual type of sample in activation analysis is a solid, the standard must be dried before use. Care must be taken if heat is used to dry the standard. Trace elements can be lost during heating when volatile compounds break down, for example if gold is present as a cyanoaurate it may be almost totally lost on heating. Addition of acid to stock solutions can help keep elements stable but if the acid is too concentrated there may be a syrupy residue left after drying. Lyophilization can be used to avoid the problem of decomposition on heating. Many of the problems of preparing chemical standards for activation analysis are described by Byrne (1984).

MULTIELEMENT STANDARDS

Since neutron activation analysis is a multielement technique it is usual to analyze for several elements simultaneously. If the samples are to be irradiated with standards then a series of containers holding single-element standards can be included but it is more convenient to combine the elements in a single multielement standard. The multielement standard saves space and speeds up the counting time afterwards.

The simplest form of multielement standard is made by combining the elements or appropriate compounds in solid form, but as with the single-element standards the concentrations required are normally low. A multielement standard can be prepared simply by doping the appropriate volume of the single-element standard solution onto a solid matrix such as filter paper, cellulose powder or silica. The single elements are added sequentially until the final result is a mixture of all the necessary elements. This is a lengthy procedure which has to be repeated for every standard.

It is more efficient to combine the individual standard solutions to form a stock solution containing all the elements of interest. It is essential to avoid cross-contamination with impurities from one standard adding to the concentration of another element of interest. Also care must be taken when combining the standards to avoid coprecipitation of incompatible elements when they are mixed in solution. This can be avoided by making up two separate solutions for the cations and anions. The problem of storing a dilute solution containing several elements is the same as for single-element solutions and however the standard is made up it should be prepared in its final form as soon as possible. The standards can be pipetted directly into the irradiation capsule and dried to avoid any losses on transfer.

The use of a solid standard avoids the problem of deterioration on storage which occurs with liquid standards. Solid solutions have been prepared as multielement standards in phenol–formaldehyde resol resin for use as a standard for biological materials (Mosulishvili et al., 1975) and geological materials (Leypunskaya et al., 1975). Twenty-one elements were incorporated into the resin and pellets weighing about 50 mg molded for use as individual standards. The homogeneity is very important when a solid standard is used and the reproducibility of the trace elements in the pellets was no worse than 6%. Metallic and organic polymer matrices have been prepared as multielement standards resulting in quite good homogeneity, with standard deviations of less than 3% (Rouchaud et al., 1987). Thirty-three elements in the form of oxides mixed with boric acid were used as a standard for photon activation analysis (Kato et al., 1976). With care it is possible to prepare a multielement standard containing

112 PREPARATION OF STANDARDS

about 40 elements for routine analysis and once prepared it can last several years. Examples of multielement standards used routinely are shown in Table 9.1.

Analysts have a tendency to avoid preparation of a standard when

Table 9.1. The concentrations of trace elements in synthetic multielement standards (in μg)

Element	Biological[a]	Geological[b]	Geological[c]
Na	2920	–	
Sc	0.0977	8.9	2.00
V	–	50.4	–
Cr	9.73	12.8	20.00
Fe	292	–	–
Co	2.92	7.3	5.00
Ni	9.73		
Cu	–	34.3	
Zn	100	–	
As			10.00
Se	4.87	–	
Br	19.6		
Rb	29.2	332	10.00
Ag	2.92	–	
Sn	48.4	–	
Sb	0.293	2.69	1.00
Cs	0.973	1.65	5.00
Ba	486	1320	99.90
La	0.292	271	10.00
Ce	0.49	538	20.00
Nd	0.973	–	10.00
Sm	–	28.5	1.00
Eu	–	2.98	1.00
Tb	0.487	1.26	1.00
Dy	–	9.69	
Yb	–	2.13	1.00
Lu	–	0.585	0.50
Hf	–	12.0	5.00
Ta	–	1.07	1.00
Au	0.00973	–	0.02
Hg	9.73		
Th	–	106.6	5.00
U	–	1.92	1.00

Sources: [a]Mosulishvili et al., 1975; [b]Leypunskaya et al., 1975; [c]Bennett, unpublished data

more than a few elements are to be analyzed. Instead they resort to certified reference materials for use as primary standards. Becker (1987) reported that of the papers published in 1984/5 on multielement analysis only 35% used primary elemental standards for all elements analyzed. The remainder relied on reference materials. Both the National Institute of Standards and Technology, US (Becker, 1987) and the International Atomic Energy Agency, Austria (IAEA, 1984) have emphasized that reference materials should not be used as primary standards. Not only are the uncertainties on recommended values too large to give accurate results but also the material is limited and should only be used for occasional quality assessment.

FLUX MONITORS

The simplest use of a flux monitor is to measure the variation between the neutron flux seen by the sample and that seen by a standard when they are irradiated at the same time. This is very important where there is large variation in the neutron flux across an irradiation site. The aim is to produce a flux correction factor which is simply the ratio of the induced activity between the two positions. The relationship is usually measured with a flux monitor consisting of a wire or foil of material such as copper, iron or cobalt. The masses of the monitors are determined precisely before irradiation. The specific activity induced in the monitor beside the sample is then measured and compared to the activity induced in a similar foil close to the standard irradiated at the same time, to give a flux correction factor.

If the data for individual standards are stored in a database it is possible to irradiate and count a sample without a standard. The database is used to deduce what the activity of a standard would have been if it was irradiated under the same set of conditions as the monitor. In this case the specific activity of the flux monitor is compared to the activity of the monitor which was irradiated at the same time as the standard in the database. This use of a database for standard activities can work well for facilities where there is a single irradiation site and counting system for the samples. Any differences in irradiation and decay periods or detector efficiency must be corrected for accurately otherwise it can be a source of errors.

Finally it is possible to use a neutron flux monitor to predict the activity induced in any radionuclide using the so-called comparator method described in Chapter 6. The monitor is used to deduce the neutron flux, and known nuclear data are used to calculate the activity of a radionuclide irradiated in that neutron flux. Flux monitors made of iron, copper or cobalt will only give a measure of the thermal neutron flux. If the true flux is to be measured a monitor which will give information about the thermal, epithermal and fast components will be more useful, particularly if the purpose is to measure the activation of a target that has resonances in the epithermal region. The measurement of k_0 values uses the true flux measurement approach and is dependent on careful evaluation of the flux taking into account the finite thickness of the target. Monitors made of zirconium are used to measure the thermal and fast flux. The k_0 values are determined for the standards initially and they are stored in a database.

There are certain difficulties associated with the use of a monitor to deduce the neutron flux precisely, regardless of the ultimate use that is made of it. The single element monitor is often used in the form of a metal foil or wire, simply because it is the easiest physical and chemical form available that may be weighed accurately and has reproducible geometry. When used in association with samples and chemical standards the foil or wire is stuck conveniently on the top, bottom or side of the irradiation capsule. A stack of samples may have a foil on the lid of each container. After irradiation the foils are removed and counted rapidly to measure the variation along the stack. There is a danger that the very presence of the foils may affect the neutron flux that the samples see, for example a thermal neutron absorber placed in such close proximity to the sample may have an effect on the thermal neutron flux seen by the sample. Certainly a foil placed on top of a polyethylene container will experience a different neutron flux to a foil placed in the center of a stack of polyethylene containers.

GEOMETRY

A standard which is used for the quantitative analysis of unknown samples must have a physical and chemical form which reproduces the sample geometry as closely as possible unless major corrections are to be made. For example it is not accurate to compare a large volume of sample with a standard consisting of a small filter paper supporting a few micrograms of the element of interest. The difference in geometry will cause errors due to neutron flux variation, differences in self-shielding, gamma ray absorption and counting geometry. These are all effects which can be corrected for, but they must be taken into account and in some ways it defeats the purpose of using a direct standard, unless the standard reproduces the geometry of the sample as closely as possible. If the activity of an irradiated sample is compared to database values for a standard it is essential that appropriate corrections are included for any neutron self-shielding, gamma ray attenuation or counting efficiency effects caused by differences in geometry.

The neutron and gamma ray absorption effects are difficult to reproduce, but they can be evaluated by "spiking" the sample with the chemical standard and comparing the result with a clean standard. Differences between the clean standard and the standard on the sample can be used, with the sample alone, to deduce the absorption effects. This method can be used to evaluate the effect of analyzing boron compounds where the thermal neutron absorption in the sample is high. Polyethylene has the opposite effect of moderating the neutrons and enhancing thermal neutron flux. It is quite difficult to spike a lump of polyethylene and so correction procedures are harder to devise. Spiking can also be used to overcome the problem of unusual geometries, for example when large samples are used.

There are a number of ways of reproducing the sample geometry using a variety of ultrapure matrices provided that they do not contribute to the trace element composition of the standard. Frequently used supports are cellulose powder, graphite and filter paper. Vegetation can be pelletized and compared to pelletized cellulose powder supporting the standard. Cellulose filter papers can be made into all manner of shapes for difficult samples. Geological samples such as powdered rock material can be reproduced using silica powder but care must be taken to avoid the introduction of trace element contamination from the silica, since even spectroscopically pure silicon dioxide from some manufacturers contains more than 10 μg kg^{-1} of many elements of interest (Ila and Frey, 1989).

REFERENCES

Becker, D. A. (1987), "Primary standards in activation analysis," *J. Radioanal. Nucl. Chem.* **113**(1), 5–18.

Byrne, A. R. (1984), "The preparation and use of chemical irradiation standards," in IAEA, *Quality Assurance in Biomedical Neutron Activation Analysis*, IAEA–TECDOC–323, International Atomic Energy Agency, Vienna, pp. 107–120.

Ila, I. and F. A. Frey (1989), "Trace element analyses of spectroscopically pure silicon dioxide by instrumental neutron activation analysis," *J. Radioanal. Nucl. Chem.*, **131**(1), 37–42.

IAEA (1984), "Quality Assurance in Biomedical Neutron Activation," *Anal. Chim. Acta,* **165**, 1–29.

Kato, T., N. Sato and N. Suzuki (1976), "Non-destructive multi-element photon-activation analysis of environmental materials," *Talanta,* **23**, 517–524.

Leypunskaya, D. I., V. I. Drynkin, B. V. Belenky, M. A. Kolomitsev, V. Yu. Dundua, and N. V. Pachulia (1975), "Synthetic multielement standards used for instrumental neutron activation analysis as rock imitations," *J. Radioanal. Chem.,* **26**, 293–304.

Mosulishvili, L. M., M. A. Kolomitsev, V. Yu. Dundua, N. I. Shonia, and O. A. Danilova (1975), "Multielement standards for instrumental neutron activation analysis of biological materials," *J. Radioanal. Chem.,* **26**, 175–188.

Rouchaud, J. C., L. Debove, M. Federoff, L. M. Mosulishvili, V. Yu. Dundua, N. E. Kharabadze, N. I. Shonia, E. Yu. Efremova, and N. V. Chikhladze (1987), "A comparison of synthetic irradiation-resistant multielement standards for activation analysis," *J. Radioanal. Nucl. Chem.,* **113**(1), 209–215.

CHAPTER

10

REFERENCE MATERIALS

A reference material is a stable, homogeneous material produced in quantity which has well-defined values for certain physical or chemical properties. Analysis of a reference material with recommended values for the concentrations of the elements of interest is one of the most valuable ways of checking the validity of the results coming from a laboratory. The values are sometimes certified by the manufacturer and almost always the material has been analyzed by independent laboratories using a variety of analytical techniques. The recommended values for the concentrations represent the accepted value, compiled from the data produced by a number of laboratories. If the analyst is to rely on such reference materials to check a technique then care must be taken over the choice of the type of material to be used. A valuable international reference material should not be used for routine work but replaced by an in-house reference standard for use on a day-to-day basis.

CHOICE OF REFERENCE MATERIAL

There are several factors to be taken into consideration when choosing a reference material. Reference materials are generally used to assess the quality of analytical data being produced during the activation analysis procedure. If the purpose of the assessment is to check the validity of data obtained on a particular set of samples, then a reference material closely resembling the samples must be chosen. If the reason for using reference materials is to assess the quality of data being produced in general, over a range of sample types, then the exact nature of the reference material is not so critical.

The most important aspect of a reference material is that it has reliable data for the element, or elements, of interest. The material will only be useful if its elemental concentrations are well-known and the errors on the values are low. It is also useful if there is a reasonable supply of the material, since it is inconvenient to have to change to a new reference material and sometimes difficult to find suitable material which is well analyzed.

117

The second most important criterion when choosing a reference material is that it should be similar in nature to the samples to be analyzed. For example rock, vegetation, biological tissue, coal fly ash all have very different physical forms which will affect the nature of the irradiation and the counting geometry. The major element compositions are not similar and so the activation products, the background effects and interferences will be quite different in each case.

The third criterion for selecting a reference material is the trace element composition of the material. This is less critical than the need to reproduce the major element composition because the activation process is directly proportional to the amount of element present over the entire range of concentrations. It is unlikely that the trace element composition of any reference material will closely resemble a sample in all respects.

To summarize, it is more important to have well-analyzed reference standards than to match the matrix and trace element composition exactly. Inferior or unreliable material should not be used simply in order to reproduce the nature of the sample precisely. In many cases there may not be material close to the samples and compromises must be sought. In activation analysis the matrix effects are less significant than is the case for other techniques, such as X-ray fluorescence, and consequently it is possible to compare similar samples but without the identical matrices. There are exceptions, as in the case of the mineral tourmaline, where the high boron content will have a significant absorption effect on the thermal neutron flux in the reactor.

As a general guide, it is important to choose a reference material which is in the same general group as the samples which are being studied. For example, a silicate matrix rock material can be used to check the determination of trace elements in soils, basalt or granite. In agricultural studies there are similarities between the reference materials prepared from orchard leaves, kale or citrus leaves. Medical reference materials include dried liver and blood, which may be sufficiently representative of clinical material to be used in the assessment of activation spectrometry techniques. Information concerning available reference materials is found in the catalogues of the relevant organizations and details are given in the following sections.

GEOCHEMICAL REFERENCE SAMPLES

The reference materials used in geochemical work include rocks, minerals, ores and concentrates. The ores and concentrates are produced to support the mining industry, often with a known amount of just one element, such as uranium, gold or fluorine, present in a relatively high concentration. These reference samples will be included in the section on industrial reference materials. The other group of samples are rocks, minerals, sediments and soils of different composition which are produced for general analysis and are characterized for all the elements that are determined by normal techniques. Samples are prepared in large amounts at a time, regardless of their type. Each sample is crushed and ground to a fine powder in order to homogenize it.

In many countries the national geological survey is responsible for producing geochemical reference samples. Table 10.1 lists the major sources of geochemical reference samples and their abbreviations. The United States Geological Survey (USGS) and the Canadian Certified Reference Material Project (CCRMP) in particular have produced a number of useful standards. The USGS runs interlaboratory comparisons and simply publishes the mean weighted values with confidence intervals. The CCRMP tends to specialize in ores and concentrates and actually produces a certified value for its material. On the other hand, institutions

Table 10.1. Sources of geochemical reference samples

Code	Source
ANRT	Association Nationale de la Recherche Technique
BAS	Bureau of Analysed Samples, UK
BCR	Community Bureau of Reference, Brussels
BCS	British Chemical Standards
CCRMP	Canadian Certified Reference Materials Project
CRPG	Centre de Recherches Petrographiques et Geochimiques
GIT–IWG	Groupe International de Travail–International Working Group
GO–ITA	Gruppo Ofioliti–Italy
GSJ	Geological Survey of Japan
IAEA	International Atomic Energy Agency, Vienna
IGB	Institute of Geology, Bulgaria
IGEM	Institute of Geology of Ore Deposits, Petrology, Mineralogy and Geochemistry, Moscow
IGI	Institute of Geochemistry, Irkutsk (USSR)
IRSID	Institut de Recherches de la Siderurgie
MINTEK	Council for Mineral Technology
NIST	National Institute of Standards and Technology
NRC	National Research Council of Canada
USGS	United States Geological Survey, Reston
ZGI	Zentrales Geologisches Institut, Berlin, Germany

like the National Institute of Standards and Technology, United States (NIST) provide recommended values for elements in their reference samples, calculated from values obtained by selected laboratories. There are many other sources of reference materials from France, Japan, Germany and South Africa for example where the geological surveys have particular sample types that are analyzed for certain elements.

Table 10.2, which lists examples of the geochemical reference samples available from a number of sources, demonstrates the large number and

Table 10.2. Geochemical reference samples

Sample type	Source
Andesite	GSJ, USGS
Anorthosite	ANRT, IGEM, GIT–IWG
Basalt	CRPG, GSJ, NIST, USGS, ZGI, IGEM, GIT–IWG
Bauxite	HUN, NIST, ANRT, IGI, BCS
Biotite	CRPG
Diabase	IGEM, USGS
Dolomitic limestone	IGI, NIST
Dunite	USGS, IGEM, MINTEK
Feldspar	IAEA, NIST, ANRT
Flint clay	NIST
Gabbro	CCRMP, IGEM, GSJ, IGI, GO–ITA
Granite	USGS, CRPG, ZGI, IGB, IGEM, ANRT, GIT–IWG, MINTEK
Granodiorite	USGS, IGEM, GSJ
Kimberlite	IGEM, MINTEK
Lake sediment	IAEA
Limestone	ZGI, IRSID, BCS
Marine sediment	IGI, NRC
Obsidian	NIST
Peridotite	IGEM, USGS
Phosphate rock	NIST, BCR
Potash feldspar	NIST, BCS
Pyroxenite	MINTEK
Quartz	MINTEK
Rhyolite	GSJ, USGS
Serpentine	ZGI, MINTEK
Soda feldspar	BCS, NIST
Soil	CCRMP, IAEA, MINTEK, IGI, USGS
Stream sediment	MINTEK, IGGE
Syenite	CCRMP, MINTEK, USGS

variety of samples types that there are. The most recent compilation of geochemical reference samples (Flanagan, 1986) has 830 entries with a list of 55 suppliers, covering a wide variety of rocks and minerals. Even a compilation of reference samples listing just silicate rocks and minerals (Govindaraju, 1984) gives the elemental compositions of 170 international reference samples.

There are a lot of data for trace elements in reference samples published in papers where the authors have determined perhaps a few elements for a particular reference sample. A recent review of the literature published in 1987 (Roelandts, 1988) shows that of 241 references to geochemical reference samples in the literature, 30% concerned the rare earth elements and next were the transition elements. At present there is little data available for the platinum group elements except in ore material.

Most geochemical reference samples have been standardized by a variety of techniques including spectrophotometry, atomic absorption spectrometry, electron microprobe analysis, X-ray fluorescence, inductively coupled plasma emission spectrometry, mass spectrometry and neutron activation analysis. Neutron activation analysis is, however, the technique providing the most information on the elemental composition of these samples. This is supported by a survey (Roelandts, 1988) showing that 58 of the 341 papers published in 1987 containing data on geochemical reference samples, appeared in *Journal of Radioanalytical and Nuclear Chemistry* and provided 2226 of the 8754 new data for elemental composition.

It is interesting to note the major contribution made by activation analysis in view of the dependence of analysts on the use of reference materials as standards. Becker (1987) reported that 65% of the papers published in *Journal of Radioanalytical and Nuclear Chemistry* during 1984 and 1985 used reference materials as standards. Since about 26% of the data published on reference samples is the result of neutron activation analysis (Roelandts, 1987) the inference is that the majority of data obtained for new geochemical reference samples using neutron activation analysis are standardized using the recommended or consensus values for other reference samples.

It is concluded that care should be exercised when using published values for geochemical reference materials. It is now the trend to publish compilations of consensus values for geochemical reference materials which use all the data available to produce a mean figure. In particular, compilations are produced for NIST, USGS and CCRMP reference samples containing consensus figures. All the data used and references to their sources are provided with the consensus value so that the reader

Table 10.3. Summary of consensus values for reference rock BHVO–1

Element	(Units)	Mean	±	SD	(n)
Ag	(ppb)	55	±	7	(5)
Al	(%)	7.3	±	0.11	(33)
As	(ppm)	0.4	±	0.22	(6)
Au	(ppb)	1.6	±	0.5	(10)
B	(ppm)	2.5	±	0.6	(8)
Ba	(ppm)	139	±	14	(37)
Be	(ppm)	1.1	±	0.3	(7)
Bi	(ppb)	18	±	4	(9)
Br	(ppm)		0.71		(2)
C	(ppm)	98	±	51	(7)
C–I	(ppm)		36		(1)
Ca	(%)	8.15	±	0.12	(32)
Cd	(ppb)	69	±	11	(5)
Ce	(ppm)	39	±	4	(56)
Cl	(ppm)	92	±	8	(12)
Co	(ppm)	45	±	2	(33)
Cr	(ppm)	289	±	22	(36)
Cs	(ppm)	0.13	±	0.06	(8)
Cu	(ppm)	136	±	6	(15)
Dy	(ppm)	5.2	±	0.3	(28)
Er	(ppm)	2.4	±	0.2	(18)
Eu	(ppm)	2.06	±	0.08	(50)
F	(ppm)	385	±	31	(11)
Fe	(%)	8.55	±	0.15	(39)
Fe_2O_3	(%)	2.82	±	0.24	(8)
FeO	(%)	8.58	±	0.09	(12)
Ga	(ppm)	21	±	2	(6)
Gd	(ppm)	6.4	±	0.5	(31)
Ge	(ppm)		1.64		(2)
H	(ppm)		−		
H_2O^+	(%)	0.16	±	0.06	(10)
H_2O^-	(%)	0.05	±	0.01	(3)
H_2O-T	(%)	0.22	±	0.07	(4)
Hf	(ppm)	4.38	±	0.22	(30)
Hg	(ppb)		5.6		(2)
Ho	(ppm)	0.99	±	0.08	(16)
I	(ppm)		−		
In	(ppb)		180		(1)
Ir	(ppb)		0.44		(1)
K	(%)	0.430	±	0.029	(37)
La	(ppm)	15.8	±	1.3	(53)
Li	(ppm)	4.6	±	1.5	(10)
Lu	(ppb)	291	±	26	(32)
Mg	(%)	4.36	±	0.13	(33)

Table 10.3. Continued

Element	(Units)	Mean	±	SD	(n)
Mn	(ppm)	1 300	±	62	(43)
Mo	(ppm)	1.02	±	0.10	(9)
N	(ppm)		22.6		(1)
Na	(%)	1.68	±	0.05	(38)
Nb	(ppm)	19	±	2	(19)
Nd	(ppm)	25.2	±	2.0	(45)
Nd–143/144			0.512 986		(2)
Ni	(ppm)	121	±	2	(29)
O	(%)		–		
Os	(ppm)		<22		
P	(ppm)	1 190	±	110	(23)
Pb	(ppm)	2.6	±	0.9	(7)
Pd	(ppb)	3.0	±	0.4	(3)
Pr	(ppm)	5.7	±	0.4	(9)
Pt	(ppb)		2.2		(1)
Rb	(ppm)	11	±	2	(27)
Re	(ppm)		<10		
Rh	(ppb)		0.2		(1)
Ru	(ppm)		<0.46		
S	(ppm)	102	±	7	(4)
Sb	(ppm)	0.159	±	0.036	(12)
Sc	(ppm)	31.8	±	1.3	(36)
Se	(ppb)	74	±	44	(6)
Si	(%)	23.32	±	0.25	(26)
Sm	(ppm)	6.2	±	0.3	(53)
Sn	(ppm)	2.1	±	0.5	(8)
Sr	(ppm)	403	±	25	(32)
Sr–87/86			0.703 48		
Ta	(ppm)	1.23	±	0.13	(26)
Tb	(ppm)	0.96	±	0.08	(35)
Te	(ppb)	6.4	±	1.6	(3)
Th	(ppm)	1.08	±	0.15	(32)
Ti	(ppm)	16 220	±	380	(31)
Tl	(ppb)	58	±	12	(5)
Tm	(ppb)	330	±	40	(16)
U	(ppm)	0.42	±	0.06	(15)
V	(ppm)	317	±	12	(26)
W	(ppm)	0.27	±	0.06	(5)
Y	(ppm)	27.6	±	1.7	(22)
Yb	(ppm)	2.02	±	0.20	(57)
Zn	(ppm)	105	±	5	(15)
Zr	(ppm)	179	±	21	(27)

Source: Gladney and Roelandts, 1988

can assess the quality of the final figure. Such compilations are published regularly in *Geostandards Newsletter* (Gladney et al., 1987). An example of such a compilation is given in Table 10.3 for a well-known reference sample, a basalt called BHVO-1 from the USGS (Gladney and Roelandts, 1988). The list of elements with their consensus values demonstrates the wide range of elements with known values and indicates the errors associated with them.

BIOLOGICAL REFERENCE MATERIALS

Biological materials represent a very much smaller group of reference materials. The term biological includes agricultural, clinical and marine material. The preparation of such samples is on the whole more rigorous than for geological material because of the very low concentrations of trace elements encountered in them. Clinical material is particularly prone to contamination during preparation. In general the samples are dried, ground and homogenized in conditions designed to minimize contamination. Sample sizes tend to be smaller than the geological samples, because of the problems involved in collection and homogenization of material such as tissue.

The majority of biological reference materials have been produced by the IAEA or by the NIST. It is the purpose of the Analytical Quality Control Services program of the IAEA "to assist laboratories engaged in the analysis of nuclear, environmental, biological, and materials of marine origin for radionuclide, major, minor and trace elements, as well as stable isotopes using atomic and nuclear analytical techniques, to check the quality of their work" (IAEA, 1989). The NIST responds to requests "from sources such as industry, government agencies, standards bodies, professional societies, trade associations and individuals." A standard reference material "is produced in response to a demonstrated measurement need." (Alvarez and Uriano, 1985). There are other sources of reference material, perhaps the most widely used being "Bowen's Kale" which was grown and homogenized by Bowen at the University of Reading (Bowen, 1985). A list of producers of reference materials and their samples are given in Table 10.4. A valuable review of recent current materials is found in a publication by Wolf (1985), which contains contributions from the NIST (Alvarez and Uriano, 1985), IAEA (Parr, 1985) and BCR (Wagstaffe, 1985).

The IAEA produces its reference values by distributing the material to as large a number of laboratories as possible, throughout the world, and averaging the data that are sent back. The result is a comprehensive

Table 10.4. Biological reference material

Sample type	Description	Source
Agricultural	Brewers yeast	NIST
	Citrus leaves	NIST
	Cellulose, cotton	IAEA
	Diet, human	IAEA
	Hay	IAEA
	Kale	Bowen
	Milk powder	IAEA, NIST
	Olive leaves	BCR
	Orchard leaves	NIST
	Rice flour	NIST
	Rye flour	IAEA
	Spinach	NIST
	Tomato leaves	NIST
	Wheat flour	NIST
	Whey powder	IAEA
Clinical	Blood, animal	IAEA
	Bone, animal	IAEA
	Muscle, animal	IAEA
	Liver, bovine	NIST
	Serum, human	NIST
Marine	Copepoda, dried	IAEA
	Fish flesh	IAEA
	Fish tissue	IAEA
	Oyster tissue	NIST

list of elements with mean values and errors. Sometimes the errors quoted are high, if the number of laboratories analyzing for a particular element is low. The NIST operates differently and works with cooperating laboratories to produce values which are certified for the elements. For the concentration of an element to be certified it is usually determined by at least two independent techniques. Other values for additional elements which are not certified are also given for information and the most recent compilation of NIST reference materials includes consensus values extracted from the literature. Although these data are not recommended values they provide a lot of additional information. A comparison of the data available for the milk powders produced by IAEA and NIST respectively, in Table 10.5, demonstrates the precision of the data from the two establishments and shows how similar the ranges are when consensus means are used.

Table 10.5. Trace element composition of dried milk powder reference samples

Element (units)	NIST (mean ± sd)	Consensus mean (n)	Range	IAEA (mean)	Range
Ag (ng g^{-1})	<0.3	<0.3			
Al (μg g^{-1})	2	<3			
As (ng g^{-1})	1.9	1.77 (1)		4.85	4.83–4.87
Br (μg g^{-1})	12	11.85 (2)	11.6–12.1		
Ca (%)	1.3 ± 0.05	1.263 (1)	1.2–1.326	1.29	1.21–1.37
Cd (ng g^{-1})	0.5 ± 0.2	0.47 (1)		1.7	1.5–1.9
Cl (%)	1.09 ± 0.02	1.085 (1)		0.908	0.734–1.082
Co (ng g^{-1})	4.1	4.12 (1)		4.5	3.7–5.3
Cr (ng g^{-1})	2.6 ± 0.7	2.5 (1)		17.7	13.9–21.5
Cu (ng g^{-1})	700 ± 100	628 (2)	606–650	378	347–409
Fe (μg g^{-1})	2.1	2.03 (2)	1.76–2.3	3.65	2.89–4.41
Hg (ng g^{-1})	0.3 ± 0.2	0.16 (1)		3.2	2.6–3.8
I (μg g^{-1})	3.38 ± 0.02	3.2 (1)		0.087	0.081–0.093
K (%)	1.69 ± 0.03	1.735 (2)	1.69–1.78	1.72	1.62–1.82
Mg (μg g^{-1})	1 200 ± 30	1 190 (1)		1 100	1 020–1 180
Mn (ng g^{-1})	260 ± 60	281.5 (2)	233–330	257	251–263
Mo (ng g^{-1})	340	332 (2)	322–342	92	83–101
Na (μg g^{-1})	4 970 ± 100	4 890 (1)		4 420	4 090–4 750
P (%)	1.05			0.91	0.808–1.010
Pb (ng g^{-1})	19 ± 3	<100		54	44–64
Rb (μg g^{-1})	11	12.75 (2)	12.4–13.1	30.8	24.5–37.1
S (μg g^{-1})	3 510 ± 50	3 514 (1)			
Sb (ng g^{-1})	0.27	0.25 (1)			
Se (ng g^{-1})	110 ± 10	100 (2)	90–110	33.9	26.7–41.1
Sn (ng g^{-1})	<500	1.9 (1)			
U (ng g^{-1})		<1			
W (ng g^{-1})		0.43 (1)			
Zn (μg g^{-1})	46.1 ± 2.2	46.75 (2)	46.6–46.9	38.9	36.6–41.2

Sources: IAEA, 1989; NIST, 1988

In summary it may be a reflection of the special skills required to prepare biological reference materials that there are far fewer available than geochemical reference samples. It is also the case that a greater use is made of activation analysis on a routine basis for geological samples and so there is less demand for biological reference samples than geochemical ones.

ENVIRONMENTAL REFERENCE MATERIALS

The requirement for environmental reference materials has occurred more recently than for geochemical or biological samples. The main producers of environmental material are limited to NIST, IAEA and the Commission of the European Communities, Community Bureau of Reference (BCR). BCR provides certified reference materials with values which are recommended by a panel of specialists on technical rather than statistical grounds. The aim of BCR is the improvement of the accuracy and comparability of measurements in areas of legislation, health, environment and commerce. Hence the reference materials are in the areas of environmental health and industrial application. IAEA samples are widely distributed but mainly to laboratories using radioanalytical techniques, there are results available for many elements in these samples and in particular for elements of interest to those using activation analysis.

The list in Table 10.6 shows that only a small range of materials is currently available as environmental reference samples. However, the major forms of sample studied in the environmental sciences are represented. Preparation of representative and homogeneous samples of media such as air particulates and water is exceedingly difficult compared to the production of a reference rock. As a result, the range of environmental reference samples available is limited at present. However, it is expected to increase as interest in environmental issues develops.

Table 10.6. Environmental reference material

Sample type	Source
Aquatic plant	BCR
Coal	NIST
Coal fly ash	BCR, NIST
Filter media	NIST
Fuel oil	NIST
Lake sediment	IAEA
Ocean water	NIST
Rainwater	NIST
River sediment	NIST
Sludge	BCR
Superphosphate fertilizer	BCR
Urban dust	NIST
Water	NIST

Table 10.7. Industrial reference materials

Sample type	Description	Source
Ores and concentrates	Antimony ore	CCRMP
	Bauxite ore	NIST
	Copper ore	BCR
	Copper concentrates	CCRMP, NIST
	Fluorspar	NIST
	Feldspar	IAEA
	Gold ore	CCRMP
	Iron ore	CCRMP, NIST
	Lead concentrate	CCRMP
	Molybdenum concentrate	NIST
	Nickel–copper–cobalt ore	CCRMP
	Noble metals bearing sulphide concentrate	CCRMP
	Precious metal bearing ore	NIM
	Tantalum ore	CCRMP
	Tin ore	BCR
	Tungsten ore	CCRMP
	Uranium–thorium ore	CCRMP
	Uranium ore	CCRMP
	Zinc concentrate	CCRMP
	Zinc ore	BCR
	Zinc–lead–tin–silver ore	CCRMP
Pure metals and alloys	Aluminum	BCR
	Copper rods	CCRMP
	Lead crystal glass	BCR
	Lead	BCR
	Molybdenum	BCR
	Nickel	BCR
	Titanium	BCR
	Zirconium	BCR
Miscellaneous	Borosilicate glass	NIST
	Coal	NIST
	Coal fly ash	NIST
	Coke	BCR
	Luxoriating oil	NIST
	Portland Cement	NIST
	Soda-lime glass	NIST

INDUSTRIAL REFERENCE MATERIALS

The number of sample types which could be prepared under the heading of industrial are almost limitless. Many of the materials used in industrial applications are very easy to prepare, consisting of pure metals and simple compounds. The interest from an analytical point of view is often limited to a few major and a few trace elements. However, because of the extremely wide variety of samples which come under the title of industrial, the demand for a particular material may be small and it may not warrant the production of a reference material. Notable exceptions occur where there is a large industry involved, for example there are a number of ferrous and non-ferrous metal reference samples, coal samples and a number of ores and concentrates. The range of reference materials and their sources is given in Table 10.7.

A significant gap occurs in the list of reference materials in the area of ultrapure non-metals, such as plastics and semiconductor material. Activation analysis is often the best method for analyzing those materials and currently there are no suitable reference materials available.

IN-HOUSE REFERENCE MATERIALS

The purpose of a certified reference material is to provide the means for a laboratory to check its techniques and test out new methods. It is not the intention of the institutions producing the material that it should be used on a regular basis for analysis. It is proposed that each laboratory should have its own primary standard made from pure chemicals to use on a day-to-day basis. The certified reference material is then used to check the method occasionally. There is therefore a place for an in-house reference material which can be used more liberally as a check on the validity of a particular run of samples. If such a secondary standard is to be used in the laboratory care must be taken to produce a homogeneous sample and in a reasonable quantity. A laboratory processing say 2000 samples per year for multielement analysis might require 10 g of standard per year, based on 0.1 g of sample every 10 samples analyzed. It is not necessary to have a wide range of standards and one geological, biological or environmental standard might be sufficient.

REFERENCES

Alvarez, R. and G. A. Uriano (1985), "New developments in NBS biological and clinical standard reference materials", in W. R. Wolf (ed.), *Biological*

130 REFERENCE MATERIALS

Reference Materials; Availability, Uses and Need for Validation of Nutrient Measurements, Wiley, New York.

Becker, D. A. (1987), "Primary standards in activation analysis," *J. Radioanal. Nucl. Chem.*, **113**(1), 5–18.

Bowen, H. J. M. (1985), "Kale as a reference material" in W. R. Wolf (ed.), *Biological Reference Materials; Availability, Uses and Need for Validation of Nutrient Measurements*, Wiley, New York.

Flanagan, F. J. (1976), "Reference samples in geology and geochemistry," *US Geological Survey Bulletin 1582*, US Government Printing Office, Washington.

Gladney, E. S. and I. Roelandts (1988), "1987 compilation of elemental concentration data for USGS BHVO–1, MAG–1, QLO–1, RGM–1, SCo–1, SDC–1, SGR–1 and STM–1," *Geostandards Newsletter*, **12**(2), 253–362.

Gladney, E. S., B. T. O'Malley, I. Roelandts, and T. E. Gills (1987), "Standard reference materials: Compilation of elemental concentration data for NBS clinical, biological, geological and environmental standard reference materials," *NBS Special Publication 260–111*, US Department of Commerce, National Bureau of Standards.

Govindaraju, K. (1984), "1984 Compilation of working values and sample description for 170 international reference samples of mainly silicate rocks and minerals," *Geostandards Newsletter, VIII, Special Issue*.

IAEA (1989), *Intercomparison Runs Reference Materials 1989*, International Atomic Energy Agency Analytical Quality Control Services, IAEA, Vienna.

NIST (1988), *NBS Standard Reference Materials Catalog 1988–89*, NBS Special Publication 260, US Government Printing Office, Washington.

Parr, R.M. (1985), "IAEA biological reference materials" in W. R. Wolf (ed.), *Biological Reference Materials; Availability, Uses and Need for Validation of Nutrient Measurements*, Wiley, New York.

Roelandts, I. (1987), "GeostandaRef Corner: Geochemical reference sample bibliography for 1986," *Geostandards Newsletter*, **11**(2), 261–278.

Roelandts, I. (1988), "GeostandaRef Corner: Geochemical reference sample bibliography for 1987," *Geostandards Newsletter*, **12**(2), 391–407.

Wagstaffe, P.J. (1985), "Development of food-oriented analytical reference materials by the Community Bureau of Reference (BCR)" in W. R. Wolf (ed.), *Biological Reference Materials; Availability, Uses and Need for Validation of Nutrient Measurements*, Wiley, New York.

Wolf, W. R. (ed.) (1985), *Biological Reference Materials; Availability, Uses and Need for Validation of Nutrient Measurements*, Wiley, New York.

CHAPTER

11

IRRADIATION TECHNIQUES

The irradiation conditions chosen for a particular measurement by activation analysis will depend on the element to be determined and the matrix. The important conditions that may be altered to enhance the sensitivity of the analysis are the nature of the neutron flux and the length of irradiation. The nuclear reactions occurring with the element of interest and the matrix will depend on the nature of the activation source. The length of the irradiation period and the length of the decay period before analysis will also affect the relative activities of the elements of interest and the matrix. All these parameters should be considered for a particular type of analysis since the activation of an element can be greatly enhanced by proper optimization of the irradiation conditions.

CHOICE OF IRRADIATION CONDITIONS

The choice of irradiation site will depend on the elements to be determined and the nature of the sample to be analyzed. Neutron activation with thermal neutrons in general provides the most useful information from a single irradiation, with the simultaneous determination of some 30–40 elements in some cases. Fast neutron activation, photon activation and the use of charged particles are all valuable for the determination of a few specific elements, which are not possible by other nuclear techniques, but as a rule they are not used in preference to thermal neutron activation for multielement analysis.

There is not usually a completely free choice of irradiation conditions in a reactor. Normally there are a limited number of irradiation sites with neutron energies over the thermal and epithermal neutron range. It may not be possible to alter the neutron flux profile except by introducing a thermal neutron filter into the irradiation site to enhance the epithermal neutron component. Where it is possible to select the neutron source and irradiation site, care should be taken to choose a highly thermalized facility for elements activated by thermal neutrons and a facility with a high epithermal neutron component to enhance elements with resonances in the epithermal region.

The length of irradiation which is attainable in practice will depend on the way in which the reactor, or alternative neutron source, is operated. The reactor may only be at full power for a fixed period each week, imposing a maximum irradiation time which may be used. An enforced decay period may result if samples can only be unloaded from the reactor at certain times of the day. At the other extreme, the operation of an irradiation facility may put a limit on the minimum time which can be measured accurately during an irradiation.

Consequently the length of irradiation and the type of flux used must be selected taking into account the facilities available. The choice of optimum conditions for one element may not coincide with those for another. Clearly a compromise has to be reached, particularly if some thirty elements are to be determined. As a general rule it is best to optimize on the elements where the sensitivity is poor, provided that the higher concentration elements can still be detected.

In some cases the conditions may not be critical at all. A long-lived radionuclide will not be greatly affected by an irradiation of one day more or less in an irradiation of 15 d. However, a radionuclide with a half-life of a few seconds cannot be irradiated in a facility where it takes 5 min to unload it. A target which is activated by thermal neutrons may not be affected significantly if it is irradiated in a neutron flux with a range of energies but the activation of a target nuclide which has resonances in the epithermal region will be greatly reduced if it is irradiated in a well-thermalized facility.

THERMAL NEUTRON ACTIVATION

Reactor irradiation sites can have neutron fluxes up to 10^{18} n m^{-2} s^{-1} and sensitivities can be quite varied depending on the irradiation facility. To give an idea of the comparison of sensitivities for different elements the best detection limits, based on a 72 h irradiation in a thermal neutron flux of 10^{18} n m^{-2} s^{-1} are given in Table 11.1. These figures are based on pure standards so they are idealized figures and they are only achievable in certain matrices such as pure silicon. The half-lives of these products vary from a few seconds to several years and an irradiation time of 72 h has been used for the calculated sensitivities.

There are a number of thermal neutron absorbers for which activation occurs solely in a thermal neutron flux. They include most of the elements below copper in the periodic table, plus several heavier elements including lanthanum, cerium, neodymium, europium, and selenium. The remaining elements that undergo neutron activation have resonances in the

Table 11.1. Thermal neutron activation detection limits for trace elements in pure silicon

Element	Nuclide	Half-life	Detection limit (μg kg^{-1})
Antimony	^{124}Sb	60.2 d	2×10^{-2}
Arsenic	^{76}As	26.3 h	1×10^{-3}
Barium	^{131}Ba	11.8 d	1.5
Bromine	^{82}Br	35.3 h	1×10^{-2}
Cadmium	^{115}Cd	44.6 d	3×10^{-1}
Cerium	^{141}Ce	32.5 d	1×10^{-1}
Cobalt	^{60}Co	5.27 y	1×10^{-1}
Chromium	^{51}Cr	27.7 d	2×10^{-2}
Cesium	^{134}Cs	2.06 y	1×10^{-2}
Copper	^{64}Cu	12.7 h	1×10^{-1}
Europium	^{152}Eu	13.4 y	2×10^{-3}
Gallium	^{72}Ga	14.1 h	5×10^{-2}
Gold	^{198}Au	2.7 d	1×10^{-4}
Hafnium	^{181}Hf	42.4 d	5×10^{-3}
Indium	^{114m}In	49.5 d	3×10^{-1}
Iridium	^{192}Ir	74.0 d	2×10^{-4}
Iron	^{59}Fe	44.5 d	4
Lanthanum	^{140}La	40.3 h	7×10^{-4}
Lutetium	^{177}Lu	6.7 d	6×10^{-4}
Mercury	^{203}Hg	46.6 d	7×10^{-1}
Molybdenum	^{99}Mo	66.0 h	1×10^{-2}
Nickel	^{58}Co	71.3 d	3
Platinum	^{199}Au	3.2 d	5×10^{-3}
Potassium	^{42}K	12.4 h	6×10^{-1}
Ruthenium	^{103}Ru	39.3 d	1×10^{-1}
Samarium	^{153}Sm	46.8 h	4×10^{-4}
Scandium	^{46}Sc	83.8 d	2×10^{-3}
Selenium	^{75}Se	119.8 h	1×10^{-1}
Silver	^{110}Ag	250 d	8×10^{-2}
Sodium	^{24}Na	15.01 h	3×10^{-1}
Strontium	^{85}Sr	50.6 d	4
Tantalum	^{182}Ta	115.4 d	1×10^{-2}
Tellurium	^{131}I	8.0 d	2×10^{-1}
Terbium	^{160}Tb	72.1 d	6×10^{-3}
Thorium	^{233}Pa	27.0 d	1.5×10^{-3}
Titanium	^{47}Sc	3.4 d	30
Tungsten	^{187}W	23.9 h	1×10^{-4}
Uranium	^{239}Np	2.4 d	2×10^{-2}
Ytterbium	^{175}Yb	4.2 d	1×10^{-3}
Zinc	^{65}Zn	243.9 d	2×10^{-1}
Zirconium	^{95}Zr	64 d	2

Source: Revel, 1987

epithermal region and are therefore enhanced by higher energy neutrons. Consequently thermal neutron absorbers are not affected by epithermal neutrons but interferences may be worse if the epithermal neutrons enhance the activation of elements in the matrix. Therefore there may be cases where it would improve the detection of one element in another, to use a thermalized flux. An example would be in the determination of aluminum using the $^{27}Al(n,\gamma)^{28}Al$ reaction, where there is an interference from the $^{28}Si(n,p)^{28}Al$ reaction induced by fast neutrons. A fully thermalized flux would result in complete removal of any of the fast neutron reactions and consequently eliminate the ^{28}Al from silicon. It is possible to have a fully thermalized neutron flux facility where the energy of the neutrons is below 0.5 eV, by placing the irradiation tube in a moderator. Since it must be well away from the neutron source, it will have a lower thermal neutron flux than an irradiation facility close to the core.

EPITHERMAL NEUTRON ACTIVATION

Epithermal neutron activation analysis is a technique which was first introduced into routine multielement neutron activation by Steinnes (1971), for the analysis of geological samples. The technique is based on the fact that some elements have isotopes with resonances in the epithermal neutron region. These targets are identified by their high resonance integrals, the epithermal equivalent of the thermal neutron cross section. A sheet of 1 mm thick cadmium will filter out all the neutrons with energies below 0.5 eV (called the cadmium cutoff energy) in a reactor spectrum. The activity induced in the total reactor neutron spectrum divided by the activity induced under a 1 mm thick cadmium filter is called the cadmium ratio. A $1/v$ absorber will have a large cadmium ratio and a nucleus with resonances will have a small cadmium ratio. Table 11.2 lists the cadmium ratios for radionuclides commonly used in activation analysis. The cadmium ratios for $1/v$ absorbers in a reactor neutron flux are generally about 30, compared to between 2 and 5 for radionuclides with resonances.

The cadmium ratio of the element of interest and that of the interfering element can be compared to decide whether it would be useful to use epithermal neutron activation to enhance the element of interest. The advantage factor (Steinnes, 1971) is a measure of the enhancement of the element of interest compared to the interfering activity from sodium. It is calculated by dividing the cadmium ratio for the interfering element by the cadmium ratio of the element of interest. A large advantage factor

Table 11.2. Cadmium ratios determined experimentally in a thermal neutron flux of 10^{16} n m^{-2} s^{-1} (cadmium ratio for gold = 2.4)

Nuclide	Cadmium ratio	Nuclide	Cadmium ratio
^{19}F	34	^{114}Cd	2.5
^{23}Na	27	^{115}In	2.4
^{26}Mg	25	^{122}Sn	3.2
^{27}Al	32	^{124}Sn	1.5
^{37}Cl	33	^{121}Sb	1.9
^{41}K	24	^{123}Sb	2.0
^{48}Ca	29	^{127}I	2.6
^{45}Sc	30	^{136}Ba	1.4
^{50}Ti	30	^{138}Ba	25
^{51}V	38	^{148}Nd	5.6
^{50}Cr	32	^{150}Nd	2.8
^{55}Mn	21	^{154}Sm	5.6
^{58}Fe	25	^{151}Eu	26
^{59}Co	12	^{160}Gd	3.5
^{58}Ni	1.0	^{164}Dy	5.5
^{63}Cu	30	^{166}Er	1.0
^{65}Cu	21	^{170}Er	5.8
^{69}Ga	2.2	^{176}Yb	2.9
^{74}Ge	9.2	^{184}W	3.6
^{75}As	2.9	^{187}Re	7.3
^{76}Se	28	^{191}Ir	8.7
^{79}Br	3.5	^{193}Ir	3.3
^{85}Rb	2.9	^{198}Pt	2.6
^{89}Y	1.3	^{197}Au	2.4
^{100}Mo	2.4	^{232}Th	3.3
^{103}Rh	4.9	^{238}U	1.5
^{108}Pd	2.2		
^{109}Ag	2.9		

Sources: Parry, 1980; 1982

indicates that the element will be greatly enhanced by epithermal activation.

The background activity in a spectrum due to a matrix component may be significantly reduced using a cadmium filter. For example the aluminum and sodium activities in rocks and in biological samples produce high backgrounds and reduce the sensitivity and precision of determination of trace elements. The improvement factor (Parry, 1980) or real advantage factor (Bem and Ryan, 1981) is a term which is used as an alternative to the advantage factor and takes into account the fact that the interference

Table 11.3. Advantage factors for thermal neutron filters

Element	Cadmium filter		Boron filter		
	Fa[a] (0.5 eV)	IF[b] (0.5 eV)	Fa[a] (10 eV)	IF[b] (5 eV)	Fs[c] (20 eV)
Aluminum				0.2	0.08
Antimony		3.1		2.1	1.7
Arsenic	14		19		1.8
Barium	1.1	4.0	1.5		6
Bromine	13	1.6	16	2.1	1.4
Cadmium	12		34		4.6
Calcium		0.2		0.07	
Chlorine			0.05	0.2	0.05
Cobalt	3.0	0.5	4.9	0.4	0.25
Copper		0.3		0.2	
Dysprosium		1		0.2	
Erbium		1		1.1	
Europium		0.2		0.3	
Gallium	9.3	2.6	15	2.6	
Germanium		0.6		0.6	
Gold	13		6.7	0.65	
Hafnium	14	2.3	9.1	2.8	
Indium	11	2.4	2	1.4	0.35
Iodine	21	2.2	25	2.9	2.2
Iridium	9.6	0.6	3.9	0.3	0.5
Lanthanum	1.4		1.9		
Lead	27		64		
Magnesium		0.2		0.2	
Manganese		0.3		0.2	0.15
Molybdenum	17	2.4	34	3.1	
Neodymium		2.0		2.8	2.8
Palladium	6.6	2.7	8.8	3.1	
Platinum	9.0	2.2	13	4.1	
Rhenium		0.8	1.3		
Rhodium	6.9	1.2	0.5		
Rubidium	19	2.00	2.7		38
Ruthenium	12				24
Samarium	6.9	1.00	1.3	0.6	8.1
Scandium		0.2	0.04		
Selenium		0.2	0.1		
Silver		2.00	1.4		
Strontium	6.2				12
Thorium	9.3	1.7	2	1.4	13
Tin	19	3.8	4.8	4.1	27

Table 11.3. Continued

Element	Cadmium filter		Boron filter		
F	Fa[a] (0.5 eV)	IF[b] (0.5 eV)	Fa[a] (10 eV)	IF[b] (5 eV)	Fs[c] (20 eV)
Titanium		0.2	0.1		
Tungsten	12	3	6.00	0.95	24
Uranium		3.8	4.7	3.8	
Vanadium		0.2	0.07	0.07	
Ytterbium		2	1.3		
Yttrium		4.4	7.9		
Zinc	6.1				10
Zirconium	27				49

Sources: [a]Advantage factor (Stuart and Ryan, 1981); [b]Improvement factor (Parry, 1984); [c]Real advantage factor (Bem and Ryan, 1981)

contributes to background in the spectrum. The improvement factor is the square root of the cadmium ratio of the interfering nuclide divided by the cadmium ratio of the element of interest. When the activity of the sample is reduced it becomes very much safer to handle and allows for the possibility of increasing the sample weight and therefore increasing the sensitivity of detection even further. It may be possible to enhance the sensitivity by counting the sample closer to the detector. Tian and Ehmann (1984) have introduced a new generalized advantage factor which includes an efficiency factor.

Cadmium is not the only absorber to be used to remove thermal neutrons. Boron cuts out neutrons at higher energies than cadmium, and boron (or a combination of boron and cadmium) can be used to optimize the effect for a particular set of interferences (Rossitto et al., 1972). A review by Alfassi (1985) describes the uses of epithermal neutron activation and compares the advantage factors derived by workers using measured cadmium and boron ratios. Examples of both factors are given in Table 11.3, calculated for sodium as the interfering element. The detection of elements with factors above unity are improved using a filter and it can be seen that a number of elements are enhanced in this way.

The decision to use cadmium or boron as the thermal neutron filter may be a compromise based on the advantage to the majority of the elements to be determined. For example, the lanthanides may be determined as a group by activation analysis. Certain elements such as samarium, terbium, ytterbium and lutetium are enhanced when epithermal activation is used to reduce sodium and scandium interferences in a rock matrix. On the other hand, lanthanum, cerium, and europium are adversely affected, having high cadmium ratios and so it may be preferable to use thermal neutron activation in order to detect these elements.

It has been shown that boron is a superior filter for some elements in biological material, while cadmium is preferred for geological material (Parry, 1984). In general the cadmium filter provides the best all-purpose filter material and a typical epithermal neutron irradiation tube installed for reactor activation analysis would be made with a cadmium lining (Holzbecher et al., 1985). However, a boron-lined irradiation tube has been installed in a reactor for the analysis of biological material (Chisela et al., 1986). The alternative to a fixed irradiation filter is to use a filtered irradiation container. It is even possible to use a container with a boron compound in its walls for irradiation in a fixed irradiation tube lined with cadmium.

FAST NEUTRON ACTIVATION

The use of 14 MeV neutron activation analysis can be helpful in the determination of a few selected elements where thermal neutron activation cannot be used. Certain elements which cannot be measured via (n,γ) reactions are possible using the (n,p), (n,α) or $(n,2n)$ reactions induced by fast neutrons. The cross sections for fast neutron reactions are generally lower than for thermal neutron activation and so the detection limits are in milligrams compared to micrograms for thermal neutrons. A compilation by McKlveen (1981) gives a comprehensive list of detection limits measured using an accelerator based on a sealed-tube generator utilizing the $^3H(d,n)^4He$ reaction. The output of the accelerator head was rated at approximately 10^{11} n s^{-1}. The detection limits taken from McKlveen's compilation are listed in Table 11.4.

Bild (1987) has recently compared the use of 14 MeV neutron activation analysis with competing techniques and indicates those elements where the technique is particularly useful. The principal examples are oxygen, nitrogen and silicon, using the reactions: $^{16}O(n,p)^{16}N$; $^{14}N(n,2n)^{13}N$; $^{28}Si(n,p)^{28}Al$, with cross sections of 39, 7.0 and 230 mb, respectively. The detection limit for oxygen is 10 μg, and it is therefore one of the most sensitive elements by this technique. The method is also very rapid since ^{16}N has a half-life of only 7.14 s, and samples may be run at a rate of 12 to 15 per hour. It is used in many applications such as the determination of oxygen in coal, iron powders, metal nitrides, rocks and niobium alloys. The gamma rays produced by ^{16}N are of high energy

(6.128 and 7.117 MeV) and so there are unlikely to be many interfering gamma rays. The main interference is a nuclear reaction of fluorine which also produces ^{16}N from the $^{19}F(n,\alpha)$ reaction.

The main advantage of fast neutron activation analysis is where the element cannot be measured using thermal or epithermal neutron activation and where the ability to analyze a solid sample non-destructively is an advantage over other analytical techniques. For example nitrogen has a detection limit of only 1 mg, which is poor compared to the alternative methods of determining nitrogen such as inert gas fusion and vacuum fusion methods. Nitrogen is also not so easy to measure because the ^{13}N produced by the (n,2n) reaction has no gamma emission. The annihilation peak at 511 keV resulting from positron emission is measured. This means there are a number of possible interfering reactions where positrons are emitted. However, because it is non-destructive the method is simple and straightforward and has been applied to the analysis of foodstuffs and fertilizer.

The application of 14 MeV neutron activation analysis to silicon is mainly in the area of geological samples. The detection limit for silicon is 0.5 mg which is adequate for the analysis of silicate rock material containing percentage levels of silicon. The main interference is the thermal neutron reaction of aluminum which can cause a problem if the aluminum content exceeds that of silicon. The main advantage of the technique is that the silicon content can be determined non-destructively prior to further neutron activation with thermal or epithermal neutrons for other elements. This technique of using the same sample is most advantageous with small precious samples such as lunar material. Other elements which have good sensitivity by 14 MeV neutron activation analysis are aluminum, antimony, barium, copper, fluorine, gallium, iron, magnesium, mercury, phosphorus, scandium, vanadium and zinc.

PHOTON ACTIVATION

The technique of photon activation is described in detail by Segebade et al. (1987). It is used mainly for the determination of light elements, namely carbon, nitrogen, oxygen and fluorine but can also applied to the determination of heavier elements to complement neutron activation analysis. The (γ,n) reactions of carbon, nitrogen, oxygen and fluorine are used most commonly, producing ^{11}C, ^{13}N, ^{15}O and ^{18}F, respectively. They are all positron emitters and do not produce gamma rays. Consequently the annihilation peak is measured at 511 keV and there

Table 11.4. Detection limits for 14 MeV fast neutron activation analysis

Element	Detection limit (μg)	Element	Detection limit (μg)	Element	Detection limit (μg)
Aluminum	900	Indium	250	Samarium	800
Antimony	550	Iodine	1 650	Scandium	1 400
Argon	Poor	Iridium	600	Selenium	1 010
Arsenic	1 400	Iron	250	Silicon	500
Barium	~1 000	Lanthanum	16 000	Silver	15
Beryllium	Poor	Lead	9 700	Sodium	4 000
Bismuth	Poor	Lithium	Poor	Strontium	100
Boron	Poor	Lutetium	1 000	Sulfur	Poor
Bromine	~1 000	Magnesium	450	Tantalum	2 200
Cadmium	230	Manganese	400	Tellurium	~10 000
Calcium	607	Mercury	2 250	Terbium	~50 000
Cerium	1 630	Molybdenum	~50 000	Thallium	4 500
Cesium	5 800	Neodymium	1 700	Thorium	~50
Chlorine	1 380	Nickel	13 450	Thulium	10 500
Chromium	3 750	Niobium	Poor	Tin	1 670
Cobalt	1 650	Nitrogen	~1 000	Titanium	9 800
Copper	90	Osmium	Poor	Tungsten	10 500
Dysprosium	600	Oxygen	10	Uranium	50
Erbium	5 000	Palladium	250	Vanadium	200
Europium	170	Phosphorus	4 200	Ytterbium	4 350
Fluorine	180	Platinum	2 400	Yttrium	1 800
Gadolinium	6 300	Potassium	21 030	Zinc	5 600
Gallium	40	Praseodymium	11 100	Zirconium	290
Germanium	~1 000	Rhenium	1 800		
Gold	170	Rhodium	~10 000		
Hafnium	100	Rubidium	600		
Holmium	1 200	Ruthenium	~10 000		

Source: McKlveen, 1981

are possible interferences. Decay curves can be plotted to determine the relative contribution from different sources or radiochemical separations can be employed.

Accelerator beam energies of about 30 MeV are used to produce good sensitivities while keeping below the threshold energy of competing nuclear reactions. Nuclear interferences which occur are, for example, $^{16}O(\gamma,\alpha n)^{11}C$. A secondary interference would be the reaction with prompt protons, deuterons or alpha particles emitted during photonuclear reactions with the sample matrix, such as $^{11}B(p,n)^{11}C$ interfering with carbon. When long irradiations are undertaken heating can occur in the target. Standardization of the irradiation between samples and standards may be difficult and beam monitors are employed by sandwiching the sample between monitors. Attenuation in the sample can be a source of error and should be evaluated. It is more satisfactory to use internal standardization with an element in the matrix of known concentration. Surface treatment after irradiation is important because contamination from atmospheric elements such as carbon, oxygen and nitrogen can occur. This problem of contamination affects all analytical techniques and in activation analysis decontamination is easier because it can be done after irradiation. Mechanical removal or chemical etching is used to clean the surface, except in the case of very short-lived products.

The heavier elements emit X-rays or gamma rays and so the method can be used instrumentally for multielement analysis. Typical detection limits are of the order of 0.01–1 μg. The technique is mainly applied to environmental problems in solids, such as air particulate samples, soil and biological material. The major advantage of photon activation over neutron activation is that sodium does not activate and therefore does not create a background problem. Lead is an example of an element which can be measured successfully with photon activation but is difficult with neutron activation. Soil has been analyzed by instrumental photon activation for Na, Mg, Cl, Ca, Sc, Ti, V, Cr, Mn, Fe, Co, Ni, Zn, As, Se, Sr, Zr, Mo, Ag, Cd, In, Sb, Cs, Ba, Hg, Tl, Pb and Bi, demonstrating the multielement capabilities of the technique (Chattopadhyay and Jervis, 1974). Fusban et al. (1982) have analyzed soil from a sewage farm for Na, Mg, Si, Cl, K, Ca, Ti, Cr, Mn, Fe, Co, Ni, Zn, Ga, Ge, As, Se, Br, Rb, Sr, Y, Zr, Nb, Mo, Ag, Cd, Sn, Sb, I, Cs, Ba, Ce, Nd, Sm, Hg, Tl, Pb and U.

CHARGED PARTICLE ACTIVATION

The use of charged particles in activation analysis is described in a comprehensive book by Vandecasteele (1988). The main areas of application are in the determination of the light elements (such as boron, carbon, nitrogen and oxygen) in metals and semiconductors, followed by the medium and heavy elements in metals and semiconductors, and finally for the analysis of environmental and geological samples. In general the method cannot be used as a multielement method of analysis like neutron activation but complements it in the determination of the light elements. Applications to samples other than metals and semiconductors are limited because of the problem of heating in the sample. For example, under vacuum the temperature in a sample is above 600° C and even when helium is used to reduce the temperature it still reaches 250° C. Consequently biological, environmental and even geological samples are badly affected by volatilization of the elements of interest or of the sample itself. The effects can be dependent on how the sample was dried prior to analysis (Xenoulis et al., 1983).

Metals and semiconductors can be seriously affected by even trace amounts of the light elements so charged particle activation provides a valuable analytical method for industrial applications. Even in the case of heavier elements if neutrons are used to analyze metals the matrix tends to become very radioactive. Charged particle activation is normally applied to bulky samples so that the thickness of the target is greater than the range of the particle in the material. The excitation function will depend on the material and the particle, so careful standardization, preferably with an identical matrix and thickness, must be used. Foils are used to monitor the beam intensity for each irradiation. The targets are prepared as discs or can be used as a powder if a special sample holder is used.

Light charged particles are used most commonly, including protons, deuterons, tritons, helium-3 and helium-4. The energies range from a few MeV to 40 MeV. The cross sections are generally low and there are a large number of interferences that can occur with competing nuclear reactions. Few of the activation products are gamma ray emitters but they are positron emitters and the annihilation peak at 511 keV can be measured. However, since most of them are positron emitters the interferences are many and decay curves may have to be plotted to isolate the contributing activities. It is therefore quite usual to have to carry out radiochemical separations for optimum detection limits. It is also usual to etch the surface of the target after irradiation to remove any surface contamination, using either chemical etching or mechanical grinding.

Boron is most commonly measured using the $^{10}B(d,n)^{11}C$ reaction, for example in the determination of boron in aluminum and aluminum–magnesium alloy (Mortier, 1984). The ^{11}C is measured after separation as carbon dioxide. Carbon is most often determined via the $^{12}C(d,n)^{13}N$ reaction, as in the case of the determination of carbon in molybdenum and tungsten (Vandecasteele and Strijckmans, 1981), and the ^{13}N was measured using gamma gamma coincidence counting. Nitrogen is usually determined via the $^{14}N(p,\alpha)^{11}C$ reaction. The carbon is measured as separated carbon dioxide, as for example in the case of the determination of nitrogen in zirconium and Zircaloy (Strijckmans and Vandecasteele, 1987). Finally, oxygen is measured using the $^{16}O(^{3}He,p)^{18}F$ reaction. An example of the use of this reaction is the determination of oxygen in tungsten, which also involves a radiochemical separation (Vandecasteele and Strijckmans, 1981).

Protons are most often used for the determination of heavier elements. There are a large number of possible interfering reactions since, in addition to the (p,n) reaction, there are (p,α), (p,αn), (p,pn) and (p, α2n) reactions induced in other elements which may result in the same activation product. The neutron produced during the (p,n) reaction can also induce an (n,γ) reaction which can interfere. Most applications of proton activation involve the analysis of pure metals such as Al, Fe, Au, Co, Ag, Ta, Nb, Dy, Ir and Rh. The good sensitivity of instrumental proton activation analysis has been demonstrated in a compilation of calculated detection limits for 10 MeV protons (Debrun et al., 1976). An interesting application of proton activation is the analysis of rhodium (Debrun et al., 1975), where the detection limits for some 27 elements are below one mg kg^{-1}. Neutron activation for multielement analysis of rhodium is difficult because the iridium in the rhodium produces long-lived high activity. Therefore in this case proton activation provides a superior technique.

LENGTH OF IRRADIATION

There are several factors to be taken into account when choosing the duration of the irradiation. There may be purely practical limitations which apply such as how quickly the results are required, the cost of the irradiation and restrictions imposed by reactor operations. Apart from these considerations it is essential to consider the growth of the radionuclide of interest and that of the other activities in the sample.

The activation curve in Figure 2.3 demonstrates that there is little gain in irradiating the sample for more than about two half-lives of the radionuclide of interest. If the half-life is long, 5 y in the case of ^{60}Co for example, then it may not be practical to irradiate the sample for even one half-life. If there are two possible targets which could be used for the determination of an element, the activation of the two nuclides should be compared by calculating the induced activity for a reasonable length of irradiation, since the saturation activity may be misleading if it takes 5 y to achieve in one case and 1 h in the other. A review of the use of short-lived radionuclides for activation analysis (Dams, 1981) shows that the majority of elements can be measured using a 1 min irradiation. Quite good sensitivities may be obtained for the elements even if the half-life of the radionuclide is quite long (2 h to 3 d) as in the case of

Mn, Ga, As, Sr, Ru, La, Eu, Ho, Lu, Os and Au. Those elements giving high count rates after such a short irradiation are Na, Rb, Sc, V, Co, Se, Rh, Ag, In, Eu, Dy, Er and Hf. However in real samples the background activity from the matrix will affect the detection limits and a comparison of the short- and long-lived radionuclides produced from the same elements (Tout and Chatt, 1980) has shown that some eight elements (Br, Ca, Cu, Dy, Hf, Rb, Sc and Se) actually have better detection limits using the short-lived radionuclide in biological, environmental and geological samples.

If the half-life of the background activity is greater than that of the radionuclide of interest then the longer it is irradiated the worse the signal to background ratio becomes. This is particularly important in the case of short-lived radionuclides where the activity can grow quite rapidly. For example, in the case of the determination of silver in a copper-bearing ore, ^{110}Ag has a half-life of 18 s and ^{66}Cu has a half-life of 5 min. An irradiation time of more than about 30 s will simply increase the copper activity while the silver activity remains fairly constant. The signal to background ratio becomes smaller and the detection of silver becomes poorer.

A longer irradiation will enhance the activity of a radionuclide with a long half-life compared to that of an interfering radionuclide with a short half-life. However in that particular case there is an even greater advantage in leaving the sample before counting to allow the short-lived radionuclide to decay away. For example, to measure the iridium content of a pure rhodium powder the sample is irradiated for several hours and left until the ^{104}Rh, with a half-life of 5 min, has decayed before counting the ^{192}Ir, which has a half-life of 74 d.

The irradiation system may be a limitation on the length of irradiation that can be used. The timing of irradiations can also be a source of difficulty, particularly in the case of irradiations lasting a minute or less. For example, irradiations of a few seconds may be subject to variations of less than a second, which still constitutes a high percentage error. It is difficult to measure the transit times for a sample capsule entering and leaving the reactor site; the time will vary with sample weight and consequently it is unlikely that the irradiation time will be known very precisely. However a system for irradiating samples should be reproducible for a fixed weight and therefore identical samples and standards can be compared. Other limitations are the restrictions imposed by the reactor operations program and the amount of radioactivity which can be safely handled by the reactor operations staff when they unload the samples.

Taking all the above considerations for multielement analysis into account, the choice of irradiation time must be a compromise to obtain

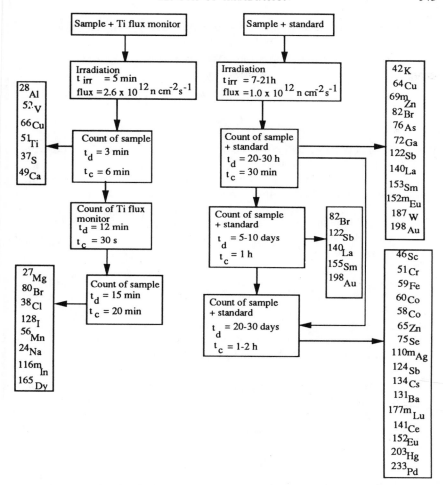

Figure 11.1. A typical irradiation and counting scheme for the analysis of air particulates, showing the schemes for both short and long-lived radionuclides. (Reproduced with permission from Dams et al., 1970.)

reasonable results for all the elements of interest. The radionuclides may be conveniently placed in two groups: those with half-lives of less than about one day and those with half-lives that are longer than a day. The figure of one day is arbitrary but it does reflect the sort of half-life that may still be considered sufficiently short that an irradiation of less than a few hours may be used for analysis. So when short irradiations are carried out radionuclides like ^{24}Na, which has a half-life of 15 h, would be measured. The best way to illustrate the separation of radionuclides into two groups is shown in Figure 11.1, which is a typical counting scheme for routine multielement analysis of air particulates (Dams et al., 1970).

REFERENCES

Alfassi, Z. B. (1985), "Epithermal neutron activation analysis," *J. Radioanal. Nucl. Chem.*, **90**(1), 151–165.

Bem, H. and D. E. Ryan (1981), "Choice of boron shield in epithermal neutron activation determinations", *Anal. Chim. Acta*, **124**, 373–380.

Bild, R. W. (1987), "Comparison of 14 MeV neutron activation analysis and competitive methods for determination of oxygen, nitrogen, silicon, fluorine and other elements," in *Comparison of Nuclear Analytical Methods with Competitive Methods*, IAEA–TECDOC–435, International Atomic Energy Agency, Vienna.

Chattopadhyay, A. and R. E. Jervis (1974), "Multielement determination in market-garden soils by instrumental photon activation analysis," *Anal. Chem.*, **46**, 1630–1639.

Chisela, F., D. Gawlik, and P. Bratter (1986), "Some problems associated with the use of boron carbide neutron filters for reactor epithermal neutron activation analysis (ENAA)," *J. Radioanal. Nucl. Chem.*, **98**(1), 133–140.

Dams, R., J. A. Robbins, K. A. Rahn and J. W. Winchester (1970), "Nondestructive neutron activation analysis of air pollution particulates," *Anal. Chem.*, **42**(8), 861–867.

Dams, R. (1981), "Selection of short-lived isotopes for activation analysis with respect to sensitivity," *J. Radioanal. Chem.*, **61**(1–2), 13–36.

Debrun, J. L., J. N. Barrandon, P. Benaben, and Ch. Rouxel (1975), "Simultaneous determination of 35 elements in rhodium samples by non-destructive activation analysis with 10 MeV protons," *Anal. Chem.*, **47**(4), 637–642.

Debrun, J. L., J. N. Barrandon, and P. Benaben (1976), "Irradiation of elements from $z=3$ to $z=42$ with 10 MeV protons and applications to activation analysis", *Anal. Chem.*, **48**(1), 167–172.

Fusban, H.-U., Ch. Segebade, and B. F. Schmitt (1982), "Instrumental multielement activation analysis of soil samples," *J. Radioanal. Chem.*, **67**(1), 101–117.

Holzbecher, J., A. Chatt, and D. E. Ryan (1985), "SLOWPOKE epi-cadmium neutron flux in activation analysis of trace elements," *Can. J. Spectr.*, **30**, 67–72.

McKlveen, J. W. (1981), *Fast Neutron Activation Analysis: An Elemental Data Base*, Ann Arbor Science, Michigan.

Mortier, R., C. Vandecasteele, K. Strijckmans, and J. Hoste (1984), "Determination of boron in aluminum and aluminum–magnesium alloy by charged particle activation analysis," *Anal. Chem.*, **56**(2), 2166–2170.

Parry, S. J. (1980), "Detection limits in epithermal neutron activation analysis of geological material," *J. Radioanal. Chem.*, **59**(2), 423–427.

Parry, S. J. (1982), "Epithermal neutron activation analysis of short-lived nuclides in geological material," *J. Radioanal. Chem.*, **72**(1–2), 195–207.

Parry, S. J. (1984), "Evaluation of boron for the epithermal neutron activation analysis of short-lived radionuclides in geological and biological matrices," *J. Radioanal. Nucl. Chem.*, **81**(1), 143–151.

Revel, G. (1987), "Present and future prospects for neutron activation analysis compared to other methods available," in IAEA, *Comparison of Nuclear Analytical Methods with Competitive Methods*, IAEA-TECDOC-435, pp. 147–162.

Rossitto, F., M. Terrani, and S. Terrani (1972), "Choice of neutron filters in activation analysis," *Nucl. Instr. Meths.*, **103**, 77–83.

Segebade, C., H-P. Weise, and G. J. Lutz (1987), *Photon Activation Analysis*, Walter de Gruyter, Berlin.

Steinnes, E. (1971), "Epithermal neutron activation analysis of geological material", in A. O. Brunfelt and E. Steinnes, (eds), *Activation Analysis in Geochemistry and Cosmochemistry*, Universitetsforlaget, Oslo, pp. 113–128.

Strijckmans, K., and C. Vandecasteele (1987), "Activation analysis with charged particles," *Anal. Chim. Acta*, **195**, 141–152.

Stuart, D. C. and D. E. Ryan (1981), "Epithermal neutron activation analysis with a SLOWPOKE nuclear reactor," *Can. J. Chem.*, **59**, 1470–1475.

Tian, W-Z. and W. D. Ehmann (1984), "Observations relative to epithermal and fast neutrons in INAA", *J. Radioanal. Nucl. Chem.*, **84**(1), 89–102.

Tout, R. E. and A. Chatt (1980), "A critical evaluation of short-lived and long-lived neutron activation products for trace element determinations", *Anal. Chim. Acta,* **118**, 341–358.

Vandecasteele, C. and K. Strijckmans (1981), "Determination of traces of carbon, nitrogen and oxygen in molybdenum and tungsten," *Talanta*, **28**, 19–23.

Vandecasteele, C. (1988), *Activation Analysis with Charged Particles*, Ellis Horwood, Chichester.

Xenoulis, A. C., A. E. Aravantinos, and C. E. Douka (1983), "The stability of biological specimens during charged particle bombardment," *J. Radioanal. Chem.*, **77**(1), 207–222.

CHAPTER

12

COUNTING TECHNIQUES

The counting conditions are chosen to optimize the sensitivity and accuracy for the determination of the element of interest. They will most probably be a compromise if a number of elements are to be analyzed simultaneously, as for the choice of irradiation conditions. First the most appropriate gamma ray must be chosen and then the detector selected to suit the energy of the gamma ray and the activity of the sample itself, taking into account the energy range, efficiency and resolution of the detector. Finally the mode of counting is selected, with the possibility of using special techniques to enhance the sensitivity, particularly in the case of short-lived radionuclides.

CHOICE OF GAMMA RAY ENERGY

The choice of a suitable gamma ray depends on which radionuclides are available. A few elements produce no radionuclides that emit gamma rays on activation and some elements produce only a single radionuclide. Most elements have just one or two suitable radionuclides and many of those have several gamma rays. The most suitable gamma ray energy will be the one which gives the best detection limit, without interferences from other radionuclides.

As a general rule the best radionuclide for a particular element is the one that provides the most abundant gamma ray. The decision will be based on the number of gammas per disintegration and also the efficiency of the detector at the gamma ray energy. For example, the 81 keV gamma ray of ^{133}Ba is less intense than the major line at 356 keV, but because of the higher efficiency of a germanium detector at the lower energy end it is the largest peak in the spectrum.

Consideration must also be given to the potential interferences. The interference could be background from other radionuclides in the sample, which will tend to be higher at the lower end of the spectrum. Then the choice would favor the higher energy peaks, where the background is lowest. In fact, since the detection limit is dependent on background, it may be that a less intense peak on a lower background is better. The other type of interference is the overlapping of adjacent peaks and if they cannot be resolved or corrected for satisfactorily then an alternative line must be sought. This is the case for the determination of iridium in chromite rock where the 316 keV line of ^{192}Ir is affected by the large peak of ^{51}Cr at 320 keV. The preferred alternative is the 468 keV line of the iridium which is free from interference and being at a higher energy has a much lower background.

If there are no suitable gamma rays, due to problems with interferences in the spectrum, it is sometimes possible to use X-rays instead (Pillay and Miller, 1969). X-rays which are used in activation analysis include those of ^{239}U, at 44 keV, and ^{233}Th, at 29 keV, which are used in the determination of uranium and thorium in geological samples (Mantel and Amiel, 1973). The niobium X-ray, which is as low as 17 keV, has been used to measure the niobium content of stainless steel, where the higher energy gamma ray line at 875 keV is swamped by the background from manganese in the sample. Cadmium can be determined with an X-ray at 50 keV and this has been used to analyze environmental samples. There are compilations available which compare gamma rays and X-rays for the determination of elements, using calculated values to produce advantage factors (Habib and Minski, 1981) or measurements made on short-lived radionuclides (Parry, 1984).

In some cases there are no suitable gamma or X-rays after irradiation but there may be a suitable prompt gamma ray that could be measured. Prompt gamma rays are emitted during neutron activation and can be detected in the usual way with a semiconductor detector while the sample is being irradiated in a neutron beam. The technique is useful in the case of certain elements such as B, Cd, Sm and Gd, providing a valuable method for the determination of boron in particular which is difficult to measure by other neutron activation techniques. A compilation of detection limits in reference samples of coal and bovine liver in Table 12.1 demonstrates the sensitivity of the technique for trace elements (Failey et al., 1979).

CHOICE OF DETECTOR

Assuming that a semiconductor gamma ray detector is chosen for the analysis, then there are choices of energy range, efficiency and resolution. It is also possible to reduce the background effects with the use of special detectors such as a Compton suppression system and coincidence counting.

The energy range required for analysis will depend on the energy of the particular radionuclide of interest and which other elements are to be included in the determination. The most commonly used semiconductor detectors, germanium crystals about 50 cm^3 volume, cover the whole of the gamma ray energy range from about 50 keV to several MeV. The lower energy cutoff is due only to the aluminum casing which attenuates emissions with energies below about 60 keV. This problem is overcome with a thin window, usually of beryllium, which is incorporated into the end-window of the detector. The detectors with thin windows can be used to detect gamma rays and X-rays down to a few keV. If the radionuclide of interest has a gamma ray line below about 100 keV it is advisable to use a detector with such a window.

The efficiency of the detector is dependent on size and generally the larger the crystal the higher the efficiency. It is usual now to use a detector with an efficiency between 10 and 20%. Although increased efficiency will enhance the detection of a gamma ray line it must be remembered that it will equally increase the background counts at the same energy. The result is that the signal to background ratio will remain the same. Since the limit of detection is calculated from the statistical variation in the background, the improvement in detection limit is a square root factor. This means that if the efficiency is increased by a factor of four the detection limit will improve by a factor of two. Therefore if the background is high the efficiency will not have a significant effect on the detection of the radionuclide of interest.

Where very low levels of activity are to be measured without interference from the matrix, a high efficiency detector can greatly improve sensitivity and throughput. For example in the case of pure carbon, where the matrix produces no measurable background, trace impurities can be measured to below mg kg^{-1} levels in most cases and if the efficiency of the detector is doubled the sensitivity is doubled. Sensitivity is often limited in the case of short-lived radionuclides since the sample can only be counted over the period when it is radioactive, which may be a very short time. Again, a high efficiency detector can be used to improve sensitivity over the counting period available. Very high efficiencies can be obtained using a well-type detector where the sample sits in the middle of the crystal and almost 4π geometry can be obtained. This type of detector is ideal for very low background counting for optimum sensitivity but is rather a specialized application and not appropriate for multielement analysis.

If the spectrum contains many peaks and there are interferences then the resolution of the detector is important. For multielement analysis a typical detector resolution is in the range 1.7–2.1 keV full width half maximum at the 1.33 MeV gamma ray line of ^{60}Co. A detector with a

Table 12.1. The detection limits for prompt gamma ray analysis of coal and bovine liver

Element	Energy (keV)	Detection limit for coal (g kg^{-1})	Detection limit for bovine liver (g kg^{-1})
H	2 223	0.012	0.033
B	477	0.000 049	0.000 067
C	1 261	27	37
	4 945	16	24
N	1 885	3.4	6.5
	5 269	2.0	3.5
	10 829	0.92	1.9
Na	472	0.28	0.13
	1 368	0.27	0.27
F	1 633	6.1	7.9
Mg	585	1.2	1.9
Al	1 779	0.45	0.53
	7 724	0.86	1.4
Si	1 273	2.4	3.2
	3 539	0.61	0.83
	4 934	1.1	1.9
P	637	1.3	2.6
S	841	0.18	0.28
	2 380	0.28	0.33
	3 221	0.59	0.64
	5 421	0.34	0.44
Cl	516	0.007 2	0.012
	758/788	0.009 3	0.015
	1 164	0.009 6	0.016
K	770	0.069	0.12
Ca	1 942	0.49	0.98
Ti	342	0.037	0.041
	1 382	0.027	0.027
V	1 433	0.025	0.035
Mn	212	0.049	0.053
	847	0.008 6	0.01
Fe	352	0.34	0.35
	7 120/7 135	0.41	0.59
	7 631/7 645	0.16	0.18
Cu	278	0.17	0.18
Zn	1 077	0.44	0.58
As	165 (+163)	0.23	0.24
Se	239	0.083	0.084

Table 12.1. Continued

Element	Energy (keV)	Detection limit for coal (g kg^{-1})	Detection limit for bovine liver (g kg^{-1})
Br	244	0.12	0.19
	315	0.44	0.59
Mo	778	0.088	0.094
Cd	558	0.000 061	0.000 093
In	162	0.014	0.014
	186	0.013	0.014
	273	0.008 1	0.007 9
Nd	697	0.007 4	0.008 2
Sm	334	0.000 027	0.000 031
Gd	182	0.000 047	0.000 053
	1 185	0.000 086	0.000 15
Pb	7 368	5.1	5.6

Source: Failey et al., 1979

resolution of 1.8 keV is usually quite adequate. The resolution at the low end of the spectrum is generally quoted for the 122 keV gamma ray line of ^{57}Co. A large germanium crystal will have a resolution of the order of 600–700 eV.

A planar germanium detector, of thickness around 25 mm has a full width half maximum between 540 and 600 eV at 122 keV. The planar detector therefore is superior to the large crystals at the low energy end, but the efficiency falls off quite quickly at the higher energies and the planar detector is not suitable for counting gamma rays with energies above about 1 MeV. In fact these days the software for the processing of spectra to separate multiple peaks using Gaussian fitting is so good that in many cases the increased resolution is not necessary. However, the planar germanium detector is still used for the determination of the lanthanides as a group because of the number of low energy gamma rays, between 60 and 150 keV, which are in fact better separated using the planar detector. Other peaks can be measured higher up the spectrum and usually the range is set as high as the ^{60}Co peaks at 1173 and 1332 keV. Another advantage of the planar detector is that the background at the low energy end of the spectrum is lower than that found in larger crystals. Because the crystal is thin, the interactions of the high energy gamma rays in the germanium that result in background do not occur.

Background activity is the most common cause of poor detection limits in multielement analysis. The background can be removed electronically in two ways: by removing the counts in the background or by collecting only the counts of interest. The first method is the basis of the Compton suppression system where the background activity produced by the Compton effect is identified from the coincident gamma ray activity collected in a ring of sodium iodide detectors. The background can be reduced by up to a factor of seven in this way. That will represent an improvement in the detection limit for the radionuclide of interest of 2.6 ($B^{1/2}$), not a substantial benefit in the routine analysis of many elements simultaneously. It is a technique that is used in the field of nuclear physics where the effect may improve the precision of the determination of nuclear data and an improvement of that order is significant.

The second technique for improving detection limits, that of coincidence counting, was used in the past when only poor resolution scintillation counters were available for activation analysis and overlapping peaks could not be resolved. Some radionuclides have emissions which are in coincidence, for example iridium where there is a triple coincidence for the decay of ^{192}Ir. The counts are only recorded when they occur in coincidence and so all the other counts are ignored. Hence there is no background from other activities, except for that due to a small, measurable random summing.

COUNTING TIME

The optimum decay period before counting will depend on the elements to be determined in a sample and on the major components causing background activity and interferences. If the radionuclide of interest has a shorter half-life than the interferences, then the sample should be counted as soon as possible. If it is longer-lived than the interferences then the detection will be better the longer the decay, until the interferences are no longer measurable. In practice the spectra are not so simple and they may consist of a number of elements of interest with different half-lives and also a number of interferences with a range of half-lives. Indeed, one radionuclide of interest may also interfere with another radionuclide of interest. Often samples are counted several times in a program of analysis. The short-lived radionuclides will be counted first before they decay away. Then the short-lived interferences will be allowed to decay before the sample is recounted for longer-lived radionuclides. The sequence of count and decay can be repeated several times to optimize the determination of radionuclides with half-lives of different length.

The length of the counting period will depend on the half-life and the activity of the radionuclide, and on the time available on the equipment. A radionuclide with a short half-life in relation to the counting time will be decaying during the counting period. This may create problems with dead time corrections and will result in deteriorating signal to background ratios during the analysis. If several samples are to be counted in succession the radionuclide in the other samples will be decaying. Consequently it would be unusual to count for more than about one half-life. To count for an additional half-life would only increase the counts in the peak by 50% and if the background is not decaying the signal to background ratio will be reduced by 25%.

It is not practical to count relatively long-lived radionuclides for one half-life. The decision of how long to count for will depend on the activity in the peak of interest and the precision required. The standard deviation of the counts is equal to the square root of the counts, so 10,000 counts will have a one sigma error of 100, giving 1% relative standard deviation on the measurement of the peak, usually an acceptable error. Finally the amount of time available on the detector and the number of samples to be measured will dictate the actual length of time available in the case of very long-lived radionuclides. As an example, Figure 12.1 shows the counting scheme for optimum multielement analysis of archaeological samples where detector time is not limited (Williams and Wall, 1990).

CYCLIC ACTIVATION ANALYSIS

Cyclic activation is a technique devised to improve the sensitivity of detection for short-lived radionuclides. It was developed for small samples, particularly of biological material such as blood, hair and nails; where the material is limited. Other examples include precious mineral samples, diamonds, small sections of valuable artifacts and paint flakes.

The saturation activity of short-lived radionuclides is reached quickly on irradiation and there is no further increase in activity with time. Similarly the counting time is dictated by the half-life since the activity of a short-lived radionuclide soon decays away and further counting will not accumulate any more counts. It is, however, possible to reirradiate the sample and count it again, once the initial activity has gone. In cyclic activation the process is repeated several times and the spectra from a number of successive counts are added together to give one final total spectrum. The gamma ray peak of interest in the total spectrum will be larger than that of the peak in the individual spectrum and so the sensitivity and precision should be improved. The technique has been reviewed by Spyrou (1981). He describes the principles of the technique

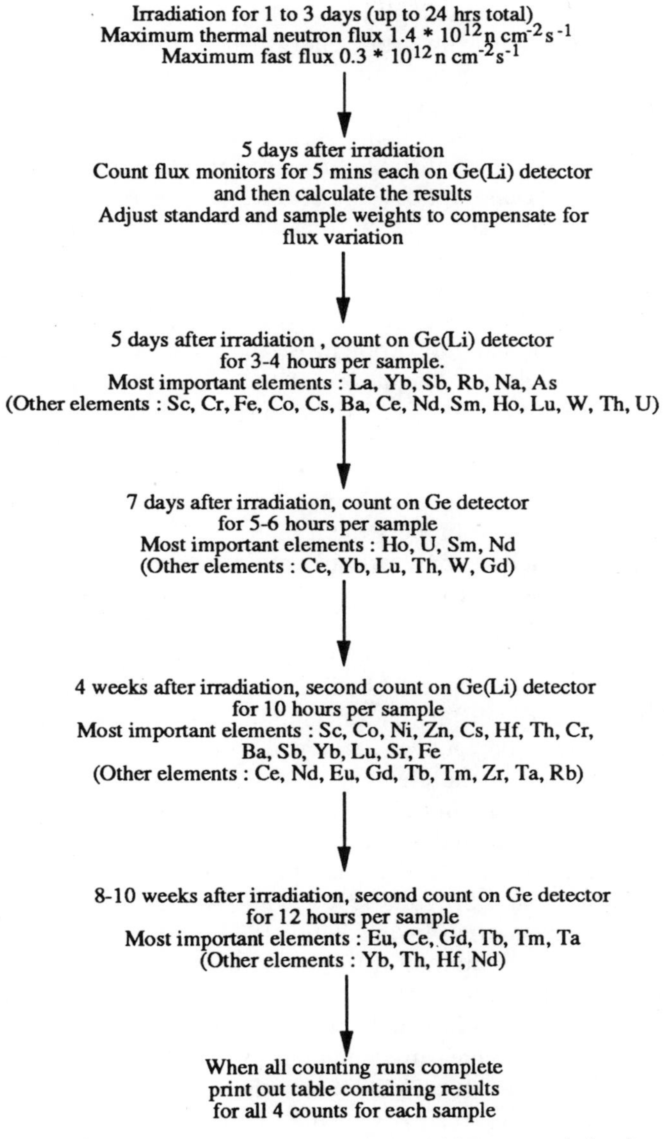

Irradiation for 1 to 3 days (up to 24 hrs total)
Maximum thermal neutron flux $1.4 * 10^{12}$ n cm^{-2} s^{-1}
Maximum fast flux $0.3 * 10^{12}$ n cm^{-2} s^{-1}

↓

5 days after irradiation
Count flux monitors for 5 mins each on Ge(Li) detector
and then calculate the results
Adjust standard and sample weights to compensate for
flux variation

↓

5 days after irradiation , count on Ge(Li) detector
for 3-4 hours per sample.
Most important elements : La, Yb, Sb, Rb, Na, As
(Other elements : Sc, Cr, Fe, Co, Cs, Ba, Ce, Nd, Sm, Ho, Lu, W, Th, U)

↓

7 days after irradiation, count on Ge detector
for 5-6 hours per sample
Most important elements : Ho, U, Sm, Nd
(Other elements : Ce, Yb, Lu, Th, W, Gd)

↓

4 weeks after irradiation, second count on Ge(Li) detector
for 10 hours per sample
Most important elements : Sc, Co, Ni, Zn, Cs, Hf, Th, Cr,
Ba, Sb, Yb, Lu, Sr, Fe
(Other elements : Ce, Nd, Eu, Gd, Tb, Tm, Zr, Ta, Rb)

↓

8-10 weeks after irradiation, second count on Ge detector
for 12 hours per sample
Most important elements : Eu, Ce, Gd, Tb, Tm, Ta
(Other elements : Yb, Th, Hf, Nd)

↓

When all counting runs complete
print out table containing results
for all 4 counts for each sample

Figure 12.1. A counting scheme to optimize the multielement analysis of archaeological and geological samples. Some radionuclides are best measured 10 weeks after irradiation. (Reproduced with permission from Williams and Wall, 1990.)

155

and some applications, including the use of cyclic activation analysis with epithermal neutrons and with a low energy photon detector.

The major drawback in cyclic activation analysis is that any radionuclides produced from elements in the matrix which have a longer half-life than the radionuclide of interest will not decay between irradiations and so the background will increase through successive irradiations. Consequently the signal to background ratio will become progressively smaller after each irradiation. The result may be that the final signal to background ratio is no better than that of a single irradiation. However, if the signal is larger it may be detectable in the total spectrum when it is not seen in the individual ones. Care must be taken therefore to choose the number of cycles so that the problem of the background activity does not begin to outweigh the advantages of the technique. Since the dead time may increase with successive irradiations, the dead time corrections must be made to the individual spectra before they are summed. With sophisticated software it is possible to carry out the process during the course of the analyses, otherwise it may be necessary to store the data for individual spectra and process them later.

The technique itself is only appropriate for very short-lived radio-nuclides, where prolonged counting will not improve the statistics. As such, the technique requires some specialized equipment since the short irradiation system has to be capable of irradiating the sample, counting it rapidly and then returning it to the irradiation site, after a decay period to allow the shorter half-life background to decay. Since the sample would usually have to be counted in the irradiation container, the capsule material must be free of the element of interest.

The technique is most suited to radionuclides with half-lives below 1 min and it has been applied successfully to ^{207m}Pb (Egan and Spyrou, 1976), ^{77m}Se (Egan et al., 1977) and the determination of fluorine in bones via ^{20}F (Kerr and Spyrou, 1978). The selenium was measured in biological reference materials by irradiating the sample for 19.5 s, allowing it to decay for 3 s, and counting it for 18 s. Seven spectra were acquired for the sample. A detection limit of about 50 μg kg^{-1} was calculated for biological material such as bovine liver or Bowen's kale. Cyclic activation analysis has been used to measure a number of trace elements in biological reference samples (Ryan et al., 1987) and the results are listed in Table 12.2 to demonstrate the sensitivity of the method.

The problem of increasing background activity may be overcome using "pseudocyclic" activation analysis (DeSilva and Chatt, 1983). In this case the decay period between cycles may be sufficiently long to allow the interfering background activity to decay away. A number of different samples are irradiated and counted and then the whole set is repeated.

Table 12.2. Determination of trace elements in reference materials with cyclic activation analysis

Element	Sensitivity (counts/µg)	Reference material	Content (mg kg^{-1})		
Ag	510	Oyster tissue	0.80	±	0.05
Au	1.6	–	–		
Br	39	Fish flour	9.3	±	0.5
Cl	37	Fish flour	0.25%	±	0.02
Dy	4 700	Oyster tissue	0.06	±	0.01
Er	160 000	–	–		
F	4.8	Fish flour	62	±	8
Ge	22	Coal	2.5	±	0.4
Hf	1 500	Urban particulate	4.1	±	0.3
In	58 000	Urban particulate	1.05	±	0.08
Pb	0.46	Urban particulate	7500	±	1500
Pd	79	–	–		
Rb	1 800	Oyster tissue	5.1	±	0.4
Sc	11 000	Oyster tissue	0.06	±	0.007
Se	1 100	Bovine liver	0.99	±	0.01
W	75	Aerosol	0.29	±	0.04
Y	6.4	Coal	0.13	±	0.02

Source: Ryan et al., 1987

This procedure can be repeated many times to build up the signal of interest. It has the advantage of allowing the interfering activity to decay away before reirradiation, while maintaining the same total analysis time for the suite of samples. The benefits of this method are discussed in a paper by Egan (1987), who compares the two techniques.

CUMULATIVE ACTIVATION ANALYSIS

The amount of sample that may be irradiated and counted for a short-lived radionuclide is limited by the activity induced in the sample. Not only would a large activated sample be a radiation hazard but the high count rate would cause problems for analyzing the sample. The sensitivity of determination of a short-lived radionuclide is then severely limited by its half-life. This problem can be overcome using cumulative activation analysis. The technique is based on irradiating and counting a number of replicates of the same sample. The single spectra are summed to give one total spectrum and the data are treated in exactly the same way as those for a single sample. Since each sample is a true replicate there is no necessity to correct spectra individually for dead time since they will all have identical dead times reflected in the final spectrum.

The technique was suggested by Guinn (1980) to improve the sensitivity of determination for short-lived radionuclides. It was applied to the determination of rhodium and silver in a reference concentrate and matte (Parry, 1982). Twenty 0.5 g samples were irradiated for 5 s in succession and the spectra summed to give a single total spectrum equivalent to a 10 g sample. The improvement over the determination of a single sample was a factor of four. More importantly the rhodium could be measured at a concentration of less than 1 mg kg^{-1}, when previously it was not detected. The precision for the determination of silver was also improved by a factor of four. The technique has also been applied to the determination of gold in ore material using the ^{197m}Au with a half-life of 7 s, using the same irradiation and counting conditions (Parry, 1987). The improvement in sensitivity is demonstrated in Figure 12.2, which shows the increase in the gold signal for successive irradiations.

The main advantage of cumulative activation analysis is that a large representative sample can be measured even in the case of material which

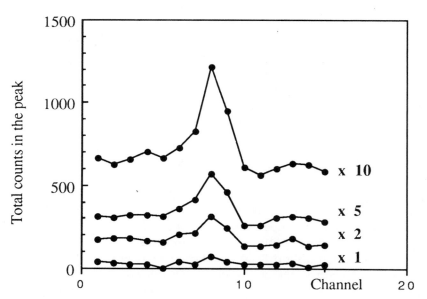

Figure 12.2. Cumulative activation of a gold ore, showing the enhancement of the ^{197m}Au peak after 10 repeated irradiation and counting cycles.

activates to form highly radioactive products. The preparation of the replicates can take more time than a single sample but the division into a number of equal portions can be done by volume. Automation of cumulative activation analysis is quite straightforward since the replicates may be treated as a series of unrelated samples, irradiated and counted in succession. Many analyzers can accommodate the continuous addition of spectra and so it is easily incorporated into a standard short irradiation facility. In the case of heterogeneous samples such as gold-bearing rocks, large samples of at least 10 g are essential for a representative analysis. In some facilities the addition of replicates is the only way that a small irradiation system can accommodate such a large sample.

DELAYED NEUTRON COUNTING

Delayed neutron counting is not a multielement method of analysis since there are few elements that produce neutrons when activated. Only uranium and thorium are measured using delayed neutrons on a regular basis and by far the majority of applications concern the determination of ^{235}U.

Delayed neutron counting systems are normally designed for the routine determination of uranium in large numbers of samples. The irradiation time and counting time for delayed neutron counting is short and, as opposed to gamma ray counting, the measurement of delayed neutrons is quite a specific technique and there are few interfering reactions. The irradiation capsule is usually made of polyethylene since the only interference is from oxygen and low density polyethylene contains very little. The sample size may be quite large and it is common to irradiate several grams of rock or up to 50 cm^3 of liquid sample. There is no special preparation of the sample although sensitivity is enhanced by the removal of water, since the oxygen present contributes to the background activity.

The choice of neutron detector will depend on the gamma radiation produced. Helium-3 counters are more sensitive to neutrons but are affected by gamma radiation to a much greater extent than the boron trifluoride tubes. A lead shield may be placed between the tubes and the sample but efficiency may be lost for neutrons by the increase in sample to detector distance. It is possible to improve the sensitivity of the tubes by increasing the gas pressure and very good improvements are achievable with pressurized boron trifluoride tubes. Efficiencies of 30–40% are commonly available in neutron counting rigs containing six or more tubes.

Delayed neutrons can have a range of energies and the configuration of the counting rig must be designed for optimum sensitivity.

The counting time is usually no longer than 1 min by which time most of the delayed neutrons have decayed. The decay time before counting is critical since some of the delayed neutrons are very short-lived. On the other hand delayed neutrons are emitted by ^{17}N produced by the fast neutron reaction with oxygen and there is a possibility of interference if, for example, water samples are being analyzed.

A typical example of a delayed neutron counting system at Los Alamos National Laboratory, US, shown in Figure 12.3, is used for a national uranium resource evaluation (Minor et al., 1982). The system is used to analyze stream sediment samples. Helium-3 detectors with a combined efficiency of about 27% are used to count the delayed neutrons. The detection limit is 0.01 μg of natural uranium per gram of sample. The system is totally automated so that a maximum of 200 samples can be loaded at one time. The delayed neutron counter is only part of a whole system which is also capable of gamma ray spectrometry and samples can be measured for both neutrons and gamma rays in sequence.

Delayed neutron analysis is used now for the determination of uranium in urine as part of the health and safety program for employees in nuclear

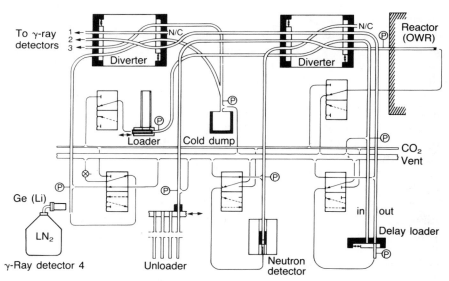

Figure 12.3. Schematic diagram of a pneumatic transfer system for delayed neutron counting and activation spectrometry. (Reproduced with permission from Minor et al., 1982.)

facilities. A high efficiency system, again at Los Alamos National Laboratory, is based on a 40% efficiency system with helium-3 detectors (Gladney et al., 1989). The thermal neutron flux is 1.2×10^{17} n m^{-2} s^{-1} and the detection limit for the determination of ^{235}U in urine is 5 μg m^{-3}, using a 25 cm^3 sample of urine, dried. A similar system at Imperial College Reactor Centre is based on pressurized boron trifluoride detectors, with an efficiency of about 30%. With a thermal neutron flux of 1×10^{16} n m^{-2} s^{-1} the detection limit for natural uranium in urine is 0.5 mg m^{-3}, using a 200 cm^3 sample, dried. The detection limit for natural uranium in soils, sediments or vegetation using this system is 100 μg kg^{-1} based on a one gram sample.The detection limit for river water is 0.5 mg m^{-3} when a 300 cm^3 sample of water is concentrated down to 30 cm^3 (Benzing et al., 1991).

REFERENCES

Benzing, R., S. J. Parry, N. M. Baghini, and J. A. Davies (1991), "Environmental and personnel monitoring for uranium by delayed neutron counting," *The Science of the Total Environment* (in press).

DeSilva, K. N. and A. Chatt (1983), "A method to improve precision and detection limits for measuring trace elements through short-lived nuclides," *J. Trace and Microprobe Tech.*, **1**(3), 307–337.

Egan, A. (1987), "Detection limits and precisions in various irradiation and counting regimes," *J. Radioanal. Nucl. Chem.*, **110**(1), 47–50.

Egan, A. and N. M. Spyrou (1976), "Detection of lead via lead-207m using cyclic activation and a modified sum-coincidence system," *Anal. Chem.*, **48**(13), 1959–1962.

Egan, A., S. A. Kerr, and M. J. Minski (1977), "Determination of selenium in biological materials using ^{77m}Se $(T =17.5\text{sec})$ and cyclic activation analysis," *Radiochem. Radioanal. Letters*, **28**(5–6), 369–378.

Failey, M. P., D. L. Anderson, W. H. Zoller, G. E. Gordon, and R. M. Lindstrom (1979), "Neutron-capture prompt γ-ray activation analysis for multielement determination in complex samples," *Anal. Chem.*, **51**(13), 2209–2221.

Gladney, E. S., W. D. Moss, M. A. Gautier, and M. G. Bell (1989), "Determination of U in urine: comparison of ICP–MS spectrometry and delayed neutron assay," *Health Physics*, **57**(1), 171–175.

Guinn, V. P. (1980), "Cyclic nuclear activation analysis," *Radiochem. Radioanal. Letters*, **44**(3), 133–138.

Habib, S. and M. J. Minski (1981), "A compilation of X- and gamma-ray sensitivities from isotopes produced by the (n,γ) reaction for utilization in instrumental neutron activation analysis," *J. Radioanal. Chem.*, **62**(1–2), 307–364.

Kerr, S. A. and N. M. Spyrou (1978), "Fluorine analysis of bone and other biological materials: a cyclic activation method," *J. Radioanal. Chem.*, **44**(1), 159–173.

Mantel, M. and S. Amiel (1973), "Simultaneous determination of uranium and thorium by instrumental neutron activation and high resolution X-ray spectrometry," *Anal. Chem.*, **45**(14), 2393–2399.

Minor, M. M., W. K. Hensley, M. M. Denton, and S. P. Garcia (1982), "An automated activation analysis system," *J. Radioanal. Chem.*, **70**(1–2), 459–471.

Parry, S. J. (1982), "Cumulative neutron activation analysis for the improved detection of short-lived nuclides," *J. Radioanal. Chem.*, **75**(1–2), 253–258.

Parry, S. J. (1984), "Neutron activation analysis with X- and low energy gamma-ray spectrometry of short-lived radionuclides," in *Proceedings of the Fifth International Conference on Nuclear Methods in Environmental and Energy Research*, University of Missouri, CONF–840408, available from NTIS.

Parry, S. J. (1987), "Cumulative neutron activation of short-lived radionuclides for the analysis of large samples in mineral exploration," *J. Radioanal. Nucl. Chem.*, **112**(2), 383–386.

Pillay, K. K. S. and W. W. Miller (1969), "Characteristic X-rays from (n,γ) products and their utilization in activation analysis," *J. Radioanal. Chem.*, **2**, 97–107.

Ryan, D. E., A. Chatt, and J. Holzbecher (1987), "Analysis for trace elements with a Slowpoke reactor," *Anal. Chim. Acta*, **200**, 89–100.

Spyrou, N. M. (1981), "Cyclic activation analysis – a review," *J. Radioanal. Chem.*, **61**(1–2), 211–242.

Williams, C. T. and F. Wall (1990), "An INAA scheme for the routine determination of 27 elements in geological and archaeological samples," *British Museum Occasional Publications* (in press).

CHAPTER

13

BIOMEDICAL APPLICATIONS

Activation analysis plays an important part in the study of trace elements in areas related to health. In clinical research there is much interest in the measurement of trace elements in the body and a compilation of elemental compositions of human tissue and body fluids is available (Iyengar et al., 1978). Certain trace elements are known to have an important role in the biochemical processes of the body and the role of trace elements in human health and disease is described by Prasad (1976). In particular, As, Ca, Cl, Co, Cu, Cr, F, Fe, I, K, Mg, Mn, Mo, Ni, Se, Si, Sn, V and Zn are considered to be essential elements; Cd, Hg and Pb may also have some physiological functions besides being toxic (Bowen, 1979). The main problem associated with determining trace elements in the body is the very low concentrations of some elements and the considerable contribution of activation analysis to the study of human health is the subject of several valuable publications (Cesareo, 1988; Cornelis et al., 1976; Heydorn, 1984). It is possible to measure the trace element composition of a person directly with *in vivo* activation analysis but usually samples of blood, tissue, urine and even hair and nails, are taken for analysis. Stable tracer techniques are used as a safe alternative to radiotracers to study the fate of trace elements in the body. The role of the diet in human health has gained importance and activation analysis has played a significant part in characterizing the trace element composition of total diets from different countries.

BLOOD AND TISSUE

The analysis of blood, tissue and organ samples for trace element composition is used to detect abnormalities which may be due to a clinical problem. Much research has been done to establish "normal" levels of the elements in humans to provide a baseline for evaluation of the levels found in sick patients. Trace element concentrations in human plasma and serum are described fully by Versieck and Cornelis (1989). The major elements found in the body are O, C, H, N, Ca, P, S, K, Na, Cl, Mg

163

and Si. Trace elements found in concentrations above 1 mg kg^{-1} are Br, Cu, Fe, F, Pb, Rb, Sr, Zn. Those with concentrations below 1 mg kg^{-1} but above 0.01 mg kg^{-1} are Al, Au, Ba, Cd, Co, Cr, Cs, I, Mn, Mo, Ni and Sn. Remaining elements are present at ultratrace concentrations in the body, below 10 μg kg^{-1}, which means that effectively they are probably below the detection limit for routine neutron activation analysis. Therefore the majority of trace elements are in very low levels in the presence of high concentrations of sodium, bromine, chlorine and phosphorus. In a review of neutron activation techniques Parr (1981) concluded that eight elements, Br, Cl, Fe, K, Na, P, Rb and Zn, can be determined non-destructively in almost all biological material and that a further group, Ca, Cd, Co, Cr, Cs, Cu, Hg, I, Mn, Sc and Se, may be determined non-destructively in some but not all biological material.

A liter of whole blood contains about 2 g of sodium and 3 g of chlorine (Iyengar et al., 1978) and the high background activity from these elements immediately after irradiation impairs the sensitivity of the determination of trace elements. Other elements such as potassium and sulphur are present in similar concentrations but do not cause major interfering activities. In the case of elements with radionuclides with half-lives longer than 15 h, the problem of sodium and chlorine activity can be overcome by letting it decay away before counting. Long-lived radionuclides are counted after several days decay and the counting times themselves can be quite long, up to 15 h per sample if very low concentrations of the elements are to be determined. For clinical research this creates no problem but as a diagnostic tool it could be limiting to have such a long delay.

A typical procedure for the routine instrumental neutron activation analysis of blood samples was used to measure 17 trace elements in 2790 samples of whole blood from the people of Uzbekistan (Zhuk et al., 1988). First the samples were lyophilized and then 20 mg samples were irradiated for 15 s in a neutron flux of 5×10^{17} n m^{-2} s^{-1} and counted for sodium and chlorine. Then a 200 mg sample was irradiated for 15 min in the same neutron flux. Potassium and bromine were measured after a decay of 4–5 d. Finally, after a decay of 20–25 d, the samples were counted for Ce, Hg, Cr, Sc, Ag, Rb, Fe, Zn and Co.

A similar procedure was used in the study of tumorous tissue (Draskovic and Bozanic, 1987). Samples weighing 0.5 g were irradiated for three days in a thermal neutron flux of 10^{17} n m^{-2} s^{-1} and then they were counted for 3000 s after a decay of 30–60 days. Michel et al. (1987) have used neutron activation to analyze tissue, organs and body fluid in the study of the interaction of cobalt–chromium alloy implants with the patient's body. Cobalt was present at levels ranging from 82 μg kg^{-1} in muscle to 41 mg kg^{-1} in the aorta. The determination of chromium at the low levels encountered in blood and tissue was difficult and chromium was only detected in the heart. The detection limits for aorta, heart, kidney, liver and spleen were 14, 23, 21, 10 and 34 μg kg^{-1}, respectively.

For elements with short-lived radionuclides, the activity from sodium and other longer-lived interferences can be removed with a radiochemical separation. Sodium and chlorine can be removed by passing the sample, in solution form, through an ion exchange column of hydrated antimony pentoxide (Girardi and Sabbioni, 1968). Further separation techniques are used where the instrumental method is affected by interferences from other longer-lived radionuclides. In the area of clinical research, where the concentrations of trace elements are very low and the samples are sometimes very small, there is considerable advantage in being able to separate an irradiated sample so there is no contamination from reagents. For example Irigaray et al. (1987) used a radiochemical separation procedure to extract arsenic from blood samples prior to analysis. Normal levels of arsenic in blood are around 1 μg dm^{-3}. In order to remove all interferences the arsenic was extracted on an ion exchange resin (DOWEX 50X8) and precipitated as trisulfure. In this particular case the work was used to look at elevated blood levels following immersion in a thermal spring containing a high concentration of arsenic.

A typical radiochemical procedure used by the National Institute of Standards and Technology (US) for biological material consists of an acid digestion followed by ion exchange separation on hydrated manganese dioxide (Greenberg et al., 1984). The elements Ag, As, Cr, Mo, Se and Sb are retained on the column. Solvent extraction with bismuth diethyldithiocarbamate in chloroform is used to extract copper from the eluate and zinc diethyldithiocarbamate in chloroform for the extraction of cadmium. Typical values for the trace elements determined in bovine liver by this technique are (in mg kg^{-1}): Ag 0.059, As 0.052, Cd 0.293, Cr 0.085, Cu 192, Mo 3.3, Sb 0.007 and Se 1.08. There has recently been a growth in interest in the determination of platinum in blood and tissue resulting from the use of cisplatinum drugs in humans. Taskaev et al. (1988) were able to measure a platinum concentration of 31 μg kg^{-1} in 250 mg samples of mouse tissue following radiochemical solvent extraction. Concentrations of platinum and gold below 100 ng kg^{-1} have been measured in bovine liver using a radiochemical precipitation method of separation (Zeisler and Greenberg, 1982).

If short-lived radionuclides are to be measured there may not be sufficient time after irradiation to carry out a radiochemical separation. Even with radionuclides which are relatively long-lived it may not be best to dissolve the sample if the element of interest is volatile, as in the case of mercury for example. Alternatively epithermal activation can be used to reduce the interfering activity from sodium, chlorine and other long-lived interferences. Studies on the different materials available for thermal neutron filters indicate that boron may be a better material than cadmium for the improvement in elements in biological material. However many of the elements of interest such as Cr, Co, Cs, Fe, Se, Sc and Zn have high cadmium ratios themselves. A practical study of the benefits of epithermal neutron activation of erythrocytes and blood plasma has shown that the detection limits using thermal neutron activation are similar or better for Br, Co, Cs, Fe, Rb, Se and Zn due to the fact that the epithermal neutron flux is generally one or two orders of magnitude down on the thermal component in a nuclear reactor (Chisela and Bratter, 1986). The elements which are improved particularly by epithermal activation, such as Dy, In, Hf, Rh, Au, U are not elements of special interest in clinical research. Another difficulty associated with epithermal neutron activation is the heating that occurs in a thermal neutron filter which has a particularly detrimental effect on medical samples. However, the one major advantage of epithermal activation analysis which was demonstrated by Chisela and Bratter (1986) was the shorter decay period required after epithermal neutron activation which meant that after 21 d decay similar results could be achieved compared to a 7 month decay period for thermal neutron activation. Details of the irradiation schemes are given in Figure 13.1.

Selenium is not enhanced by epithermal neutron activation but cyclic activation has been used to improve the detection of selenium in reference animal blood and bovine liver (Egan et al., 1977). With seven cycles of 20 s irradiation followed by an 18 s count the detection limits for ^{77m}Se were of the order of 45–65 $\mu g\ kg^{-1}$ for 200 mg of dried sample.

Activation techniques other than neutron activation have been used to analyze biological material. Charged particles can be used provided that the heating of the sample does not deteriorate it. Proton activation has been used to measure 0.0179 mg kg^{-1} of chromium in blood (Cantone et al., 1987) with the $^{52}Cr(p,n)^{52}Mn$ reaction. Cantone et al. (1985) used proton activation to measure titanium and cadmium in blood at concentrations of 0.090 g m^{-3} and 0.025 g m^{-3}, respectively. A recent comparison of neutron activation, photon activation and prompt gamma ray analysis showed that the techniques are complementary but activation with neutrons is the most efficient way of determining the essential trace elements Na, Cl, K, Ca, Cr, Fe, Zn, Se, Br, Rb and Cs (Gokmen et al., 1987). Proton-induced X-ray emission is also a complementary technique which provides data for some elements not detected by neutron activation. Papassotitiou et al. (1987) used the technique to measure K, Ca, Cr, Mn,

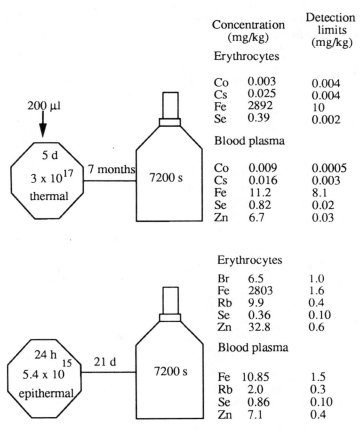

Figure 13.1. Neutron activation analysis of human blood. (Adapted with permission from Chisela and Bratter, 1986.)

Fe, Ni, Cu, Zn, Br, Rb and particularly Pb in dried skin and serum. The detection limits for lead were 0.3 mg kg^{-1} for skin and 0.1 mg kg^{-1} for serum.

STABLE TRACERS

Tracer techniques are used in clinical research to study the fate of trace elements in the body. For example a patient's blood volume can be determined by introducing the radiotracer ^{51}Cr into the bloodstream and sampling a small volume once it has been circulated in the body. Radioactive tracers are used widely in hospitals but cannot be applied to certain categories of patient, for example pregnant women and children. Provided that an element has at least two stable isotopes it is possible to use a safe alternative tracer method. A stable tracer is introduced as a replacement to the normal isotopic composition material and later sampled and analyzed by activation analysis or mass spectrometry to determine the resulting concentration.

Elements that have suitable isotopes for stable tracer studies in clinical research include Ca, Cr, Fe, Se, and Zn. An example of stable tracer techniques in the study of blood is the work of Zeisler and Young (1987) who used stable ^{50}Cr as an alternative to the radiotracer ^{51}Cr. There is an interest in the relationship between insufficient blood volume during pregnancy and infant mortality, but of course radioactive tracers cannot be used in pregnant women. Therefore the ^{50}Cr is used to measure the blood volume by activation after sampling.

Tracer studies can also be used to follow trace elements in the body. For example the absorption of iron across the gut wall can be measured by feeding a patient with radioactive ^{59}Fe and analyzing the feces after digestion. Whitley et al. (1987) used stable ^{58}Fe and ^{65}Cu tracers to study the absorption of minerals by term and preterm babies from infant formula based on cow's milk and soya protein. In another study ^{58}Fe was added as a stable tracer to a malted cocoa drink to compare the absorption of iron from ferric orthophosphate and ferrous sulfate and to see what effect adding ascorbic acid to the drink had on the absorption of iron (Fairweather-Tait et al., 1983).

HAIR AND NAILS

Since hair and nail samples are easily and painlessly obtained compared to blood or even urine samples they serve a useful purpose as an indication of trace element burden in the body. Unlike blood and tissue which indicate the essential elements, hair and nails show a higher level of those elements which are rejected by the body. Consequently analysis of hair and nails is often used in industrial health studies to detect any absorption of toxic elements such as mercury, cadmium, lead, selenium, antimony and arsenic. A full description of trace element composition, sampling problems and analytical techniques are given in a publication on human hair (Valkovic, 1988).

Hair presents a more simple matrix to analyze than blood or tissue since there is only around 1 g of sodium present in 1 kg dry weight of hair. The main problem with hair is the external contamination which must be removed before analysis. The fact that hair has a simple matrix means that it is possible to determine short-lived radionuclides of Na, Cl, Ca, Mg, Al, Mn and Cu. A routine procedure for hair analysis is shown in Figure 13.2 (Chatt et al., 1985). The combination of short and long irradiations results in the determination of Ag, Al, As, Au, Ba, Br, Ca, Cl, Co, Cr, Cu, Fe, Hg, I, K, La, Mg, Mn, Na, Rb, S, Sb, Sc, Se, Ti, U, V and Zn.

Selenium has been measured in hair samples weighing only 21–340 mg using a 10 s irradiation in a thermal neutron flux of 1.5×10^{16} n m^{-2} s^{-1} (Ohta et al., 1987). The samples are counted immediately for ^{76}Se(n,γ)^{77m}Se and values of 0.3–0.6 mg kg^{-1} were detected. Kvicala and Havelka (1988) applied the same irradiation conditions to hair samples as those used to analyze blood samples. About 200 mg of sample was irradiated for 40 h at 5×10^{17} n m^{-2} s^{-1} and counted after a 5 week decay for 15 h per sample. Another example is the analysis for toxic elements in the hair of mother and baby (Obrusnik et al., 1985). The samples were irradiated for 3 h and counted after 2–3 d decay for As, Au, Br, Cu, K, La, Na. After 20 d decay the samples were recounted for Co, Hg, Sb, Se, Zn.

Hair is often analyzed for mercury where the detection limit can be very low. An instrumental method for the analysis of beard hair is described by Pritchard and Saied (1986). They irradiated about 0.25 g of hair from dentists' beards using thermal neutron activation with a flux of 10^{16} n m^{-2} s^{-1}. After a decay of at least a week the sample was counted until there were 3000 counts in the peak of ^{203}Hg at 279 keV. Normal levels of mercury are between 0.1 and 1.5 mg kg^{-1}, dentists' beards contained more than 2 mg kg^{-1} in some cases. Noguchi et al. (1985) analyzed dentists' hair using a 1 h irradiation at 2×10^{17} n m^{-2} s^{-1}. The samples were counted after a 1-month decay for 4000–10,000 s. The detection limit was around 5 mg kg^{-1}. Mercury was measured in the hair and fingernails of dentists by Mazzilli and Munita (1986), who found that the fingernails had elevated levels of mercury although the hair did not have. Yang et al. (1985) measured both mercury and arsenic in hair and nails as exposure indicators, using instrumental neutron activation analysis.

Hair is also analyzed in forensic work as well as for medical reasons. The topic has been reviewed by Cornelis et al. (1976) who consider that the use of individual hair analyses to identify the source is not reliable because of the variable nature of hair itself. External contamination of hair was used as forensic evidence in a bullion robbery case (Kishi, 1987). Workers who handle gold bars have been shown to have a high level of gold on their hair. Activation analysis showed that the suspect had no

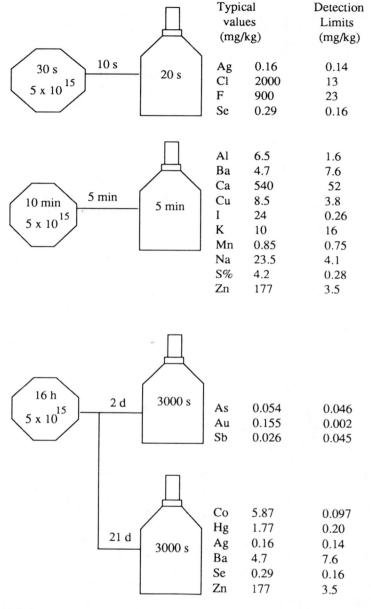

	Typical values (mg/kg)	Detection Limits (mg/kg)
Ag	0.16	0.14
Cl	2000	13
F	900	23
Se	0.29	0.16
Al	6.5	1.6
Ba	4.7	7.6
Ca	540	52
Cu	8.5	3.8
I	24	0.26
K	10	16
Mn	0.85	0.75
Na	23.5	4.1
S%	4.2	0.28
Zn	177	3.5
As	0.054	0.046
Au	0.155	0.002
Sb	0.026	0.045
Co	5.87	0.097
Hg	1.77	0.20
Ag	0.16	0.14
Ba	4.7	7.6
Se	0.29	0.16
Zn	177	3.5

Figure 13.2. Neutron activation analysis of hair. (Adapted with permission from Chatt et al., 1985.)

gold on his hair, and was later proved to be innocent. Hair was also analyzed for heavy metals to detect occupational exposure to a non-ferrous smelter (Tomza et al., 1983). The trace element composition of exposed workers was compared to controls using instrumental neutron activation analysis. The elements detected were Na, Mg, Al, Cl, K, Ca, V, Cr, Mn, Fe, Co, Cu, Zn, As, Se, Br, Ag, Cd, In, Sb, I, La, W and Au. Greatly elevated levels of As, Se, Ag, Cd and Sb were measured in the exposed group.

BONES

Fluorine is present almost exclusively as skeletal fluoride in the body and therefore it is of interest to measure the fluorine content of bone for clinical reasons. It is also, incidentally, of interest to archaeologists since during the fossilization process percentage levels of fluorine can be introduced into ancient bone material. Cyclic activation analysis has been used to measure fluorine in bone samples (Kerr and Spyrou, 1978). A 10 s irradiation followed by a 10 s count for ^{20}F was repeated 14 times and the spectra summed. The detection limit for fluorine in reference calcined bone was 52 mg kg^{-1} for a 100 mg sample, which is well below the level expected in human bone. The main limitation to detection is the nuclear interference from sodium due to the ^{23}Na(n,α)^{20}F reaction.

A recent comparison of archaeological bones and modern bones used neutron activation analysis to determine a number of major and trace elements (Hancock et al., 1987). The purpose of the work was to see if trace elements in bone could be used as an indication of diet. Samples weighing 100–600 mg were first cleaned with a carborundum abrading bit before irradiation in a flux of 10^{15} n m^{-2} s^{-1} for 1–3 min to measure Na, Mg, Al, P, Cl, Ca, Ti, V, Mn, Br, Sr, Ba, Dy and U. After 1 h decay the samples were recounted for Dy, Ba, Sr, Mn and Na. A second irradiation under cadmium was used to determine the contribution from phosphorus to aluminum and correct the figure accordingly. Although the analytical work was successful, the results indicated that the usefulness of the analysis for all trace elements in archaeological bone was extremely suspect due to the wide variations within bone samples.

URINE AND FECES

Urine and feces are usually analyzed for industrial health reasons. Urine in particular is easily sampled on a regular basis and can be analyzed by various techniques where liquid samples are readily accommodated. Activation analysis is usually seen as being a difficult technique for liquids, but if the irradiation facility can accommodate liquids or if the urine can be dried, there is no reason why it cannot be measured. However, the sodium and chlorine content of urine is very high and routine determination of trace elements by instrumental techniques with gamma ray spectrometry is as difficult as it is for blood samples.

Delayed neutron counting is not affected by non-fissionable material and so it can be used to determine uranium in urine on a routine basis. The Los Alamos National Laboratory (US) uses delayed neutron assay for screening employees who may be exposed to uranium during their work (Gladney et al., 1989). Samples up to 25 cm^3 in volume are dried and packaged in polyethylene vials before irradiation for 60 s in a thermal neutron flux of 1.2×10^{17} n m^{-2} s^{-1}. The delayed neutrons emitted by the sample are measured after a 30 s decay in a ^{3}He counting rig with an efficiency of 40%. The detection limit for uranium by this method is dependent on the isotopic ratio of the uranium, enriched uranium can be detected at 1 kBq m^{-3} but the detection limit for depleted uranium is 4 mg m^{-3}.

Instrumental neutron activation has been used for the determination of mercury in urine, to reduce the losses of volatile components during analysis (Greenberg et al., 1984). The urine was encapsulated in a quartz ampoule and sealed while cooled with liquid nitrogen. The sample was irradiated for one hour at 5×10^{17} n m^{-2} s^{-1}. The sealed sample was counted after 5 d for ^{197}Hg and after 20 d for ^{203}Hg. No losses were observed but mercury was found to be in the gaseous phase above the liquid in the quartz ampoule which required adjustments for counting geometry. The concentration of mercury measured in a reference standard of normal level material was 1.3 μg kg^{-1}.

The trace element composition of the urine of diabetics was analyzed by Ishii et al. (1980) following ashing with nitric acid. The ashes were irradiated with 5×10^{15} n m^{-2} s^{-1} for 6 or 24 h. Values were determined for Na, Mg, Al, Cl, K, Sc, V, Cr, Fe, Co, Zn, Se, Br, Rb, Sb and Cs. The urine samples from the diabetics were deficient in Cr and Se and above normal for Co.

IN VIVO ANALYSIS

It is usual to examine the elemental composition of a patient by taking samples such as blood or tissue. However sometimes the total body composition is of interest and *in vivo* activation analysis has been used for many years to determine major elements such as C, Ca, Cl, H, K, N, Na, O and P in the whole body or in an individual limb. The body is irradiated with a neutron beam from a reactor, a particle accelerator, a Van de Graaff generator or a californium source. Usually the prompt gamma rays emitted during irradiation are measured, although in some cases the delayed gamma rays are detected. The techniques and recent applications of *in vivo* activation analysis have been reviewed (Ettinger, 1988; Parr, 1981).

In vivo activation analysis has been used for the determination of nitrogen in the body. The whole body protein can be deduced from the measurement, to aid the understanding of many diseases such as obesity, anorexia, cancer, kidney and heart diseases. Larsson et al. (1987) used a ^{252}Cf source to induce the thermal neutron induced reaction $^{14}N(n,\gamma)^{15}N$ and measure the prompt 10.8 MeV neutrons that are emitted. The mean dose rate for such an experiment is 0.25 mSv during a 30 min exposure. *In vivo* activation analysis has been used in the study of renal osteodystrophy (Krishnan et al., 1987). In this disease bone mass changes non-uniformly and bone in the hand is lost progressively. This loss of bone is monitored by measuring the calcium in the hand. The hand is irradiated for 5 min with two 100 μg californium sources to induce the $^{48}Ca(n,\gamma)^{49}Ca$ reaction and the dose to the patient's hand is 75 mSv. The hand is simply counted with two sodium iodide detectors for 20 min after a 2 min decay.

Minor elements such as I, Fe and Mg and trace elements such as B, Cd, Hg, Li and Si can be measured *in vivo*. Cadmium can be determined in liver and kidney with neutron activation and prompt gamma ray analysis using the $^{113}Cd(n,\gamma)^{114}Cd$ reaction. Industrial exposure to cadmium can be measured with detection limits of 6.5 mg kg^{-1} for liver measurements and 6.4 mg for a kidney, with a skin dose of only 0.5 mSv (Franklin et al., 1987). Another example of *in vivo* activation for trace elements is a screening method for measuring the silicon in the lungs of occupationally exposed workers (Kacperek et al., 1987). A 2 MV Van de Graaff generator is used to produce a pulsed beam of 5.2 MeV neutrons. This is used to induce the fast inelastic scattering reaction $^{28}Si(n,n',\gamma)^{28}Si$. The dose to the worker is 10 mSv. The amount of silicon typically found in the lungs of the workers is 2–3 g, well above the detection limit of about 0.35 g, which is the sort of concentration found in the normal unexposed population.

NUTRITION

In recent years the significance of diet in the area of human health has gained recognition. The role of trace elements in human and animal nutrition is described in a book by Underwood (1977). The International Atomic Energy Agency has a program on the development of a total diet reference material. The original form of the mixed diet is wet and so it has been freeze-dried and homogenized. A comparison of the neutron activation method with that of other techniques (Parr, 1988) shows that the method is highly favorable for As, Cd, Co, Cr, Cu, Fe, Hg, I, Mn, Mo, Na, Sb, Se, Sn, V and Zn. It is perhaps surprising that activation analysis as a technique has played an important role in the evaluation of the reference diet. On irradiation, a matrix very high in Br, K, Na and P produces interferences which are a major problem. These can be removed using hydrated manganese dioxide ion exchange resin and the detection limits are improved by a factor of 100–2000 compared to instrumental techniques. Using a flux of 3–5×10^{17} n m^{-2} s^{-1} and a 300 mg sample the detection limits were 0.1–0.3 μg kg^{-1} for As and Sb; 1–10 μg kg^{-1} for Cr, Mo and Se.

The determination of a total daily diet is limited by high salt content but the analyses of the separate components of the diet such as dairy products, meat, cereal, vegetables and fruit, are not so difficult. The procedure used by the US Food and Drug Administration to analyze a range of food products, shown in Figure 13.3, demonstrates the capabilities of neutron activation when it is optimized (Cunningham and Stroube Jr, 1987). The freeze-dried samples, weighing about 300 mg, are irradiated for just 15 s in a thermal neutron flux of 4.9×10^{17} n m^{-2} s^{-1} before counting for Ca, Cl, Cu, Ga, I, In, K, Mg, Mn, Na, S, Sr, Tl and V. A longer irradiation of 4–5 h in the same flux was used to measure Ag, As, Au, Ba, Br, Cd, Ce, Co, Cr, Cs, Eu, Fe, Hf, La, Lu, Mo, Rb, Sb, Sc, Se, Sm, Ta, W, Yb and Zn. Not all these elements were detected in the samples and the elements determined in meat, poultry and fish are listed in Figure 13.3. Foodstuffs such as fat, sugar and beverages were irradiated in a lower neutron flux (1.3×10^{17} n m^{-2} s^{-1}) because the heating in the higher flux position resulted in leakages due to the food outgassing.

Gharib et al. (1985) analyzed dried milk powder and compared atomic absorption spectrometry and proton induced X-ray emission with neutron activation. Apart from proton induced X-ray emission, which suffered from inconsistencies, the techniques were comparable. The neutron activation procedure consisted of a 3 min irradiation for Al and V; 20 min for Mn, Mg, Cl, Na and K; 30 h for Cr, Co, Fe, Rb, Sb, Zn. Milk

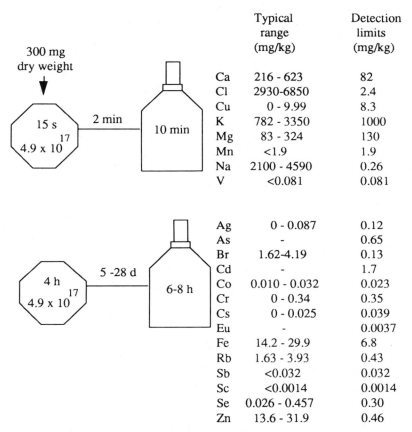

	Typical range (mg/kg)	Detection limits (mg/kg)
Ca	216 - 623	82
Cl	2930-6850	2.4
Cu	0 - 9.99	8.3
K	782 - 3350	1000
Mg	83 - 324	130
Mn	<1.9	1.9
Na	2100 - 4590	0.26
V	<0.081	0.081
Ag	0 - 0.087	0.12
As	-	0.65
Br	1.62-4.19	0.13
Cd	-	1.7
Co	0.010 - 0.032	0.023
Cr	0 - 0.34	0.35
Cs	0 - 0.025	0.039
Eu	-	0.0037
Fe	14.2 - 29.9	6.8
Rb	1.63 - 3.93	0.43
Sb	<0.032	0.032
Sc	<0.0014	0.0014
Se	0.026 - 0.457	0.30
Zn	13.6 - 31.9	0.46

Figure 13.3. Neutron activation analysis of meat, fish and poultry. (Adapted with permission from Cunningham and Stroube Jr, 1987.)

has a favorable matrix for neutron activation analysis and so detection of trace elements is quite straightforward. Seventeen trace elements have been determined in babyfoods with neutron activation using the irradiation scheme shown in Figure 13.4 (Asubiojo and Iskander, 1988). Samples, including infant milk formula, commercial cereal formula and corn were analyzed to ensure that they contained adequate concentrations of the essential elements and did not contain excessive amounts of the toxic ones. Samples weighing 250 mg were irradiated for 5 min for the short-lived radionuclides and 8 h for the long-lived radionuclides in a thermal neutron flux of 2×10^{16} n m^{-2} s^{-1}. Typical values for the milk formulas

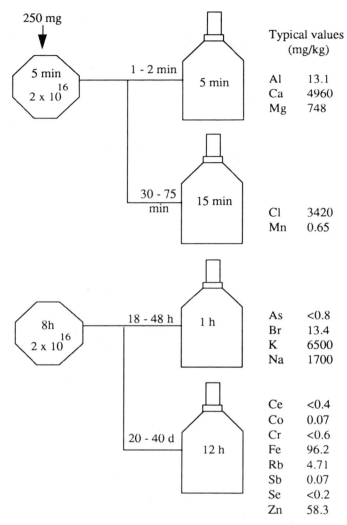

Figure 13.4. Neutron activation analysis of infant formula milk. (Adapted with permission from Asubiojo and Iskander, 1988.)

for the elements determined are given in Figure 13.4. The study actually showed that the daily supply of elements in the milk formula was adequate but that the cereal analyzed was deficient in iron and manganese.

Epithermal neutron activation was shown to improve the detection of iodine in food (Stroube Jr et al., 1987). Concentrations ranging from below 0.003 up to 0.74 mg kg^{-1} were measured in 17 different food matrices. Cyclic activation was used to determine selenium in dried food (McDowell et al., 1987). Samples weighing 250–500 mg were irradiated for 20 s, followed by a 10 s decay and a 20 s counting period. The number of cycles were limited to four by the buildup of sodium, chlorine and aluminum activity in the matrix. The detection limits were 3–5 μg kg^{-1} in desserts but only 30–40 μg kg^{-1} in meat and fish due to their high salt content.

REFERENCES

Asubiojo, O. I. and F. Y. Iskander (1988), "A trace element study of commercial infant milk and cereal formulas," *J. Radioanal. Nucl. Chem.*, **125**(2), 265–270.

Bowen, H. J. M. (1979), *Environmental Chemistry of the Elements*, Academic Press, New York.

Cantone, M. C., N. Molho, and L. Pirola (1985), "Cadmium and titanium in human serum determined by proton nuclear activation," *J. Radioanal. Nucl. Chem.*, **91**(1), 197–203.

Cantone, M.C., G. Gambarini, N. Molho, and L. Pirola (1987), "Chromium in human serum determined by proton activation," *J. Radioanal. Nucl. Chem.*, **111**(2), 351–357.

Cesareo, R. (ed.) (1988), *Nuclear Analytical Techniques in Medicine*, Elsevier, Amsterdam.

Chatt, A., M. Sayjad, K. N. De Silva and C. A. Secord (1985), "Human scalp hair as an epidemiologic monitor of environmental exposure to elemental pollutants," in IAEA, *Health-related Monitoring of Trace Element Pollutants using Nuclear Techniques*, IAEA TECDOC-330, IAEA, Vienna, pp. 33–49.

Chisela, F. and P. Bratter (1986), "Determination of trace elements in biological materials by instrumental epithermal neutron activation analysis," *Anal. Chim. Acta*, **188**, 85–94.

Cornelis, R., J. Hoste, A. Speecke, C. Vandecasteele, J. Versieck, and R. Gijbels (1976), "Activation analysis – Part 2," in T. S. West (ed.), *Physical Chemistry Series Two, International Review of Science*, Butterworths, London, pp. 106–117.

Cunningham, W. C. and W. B. Stroube Jr (1987), "Application of an instrumental neutron activation analysis procedure to analysis of food," *The Science of the Total Environment*, **63**, 29–43.

Draskovic, R.J. and M. Bozanic (1987), "Statistical investigations of some element distributions in healthy and pathologically altered human colon mucosa. 1. Distribution of some elements in healthy mucosa and mucosa with inflammation or tumorous processes," *J. Radioanal. Nucl. Chem.*, **116**(2), 409–420.

Egan, A., S. A. Kerr, and M. J. Minski (1977), "Determination of selenium in biological materials using ^{77m}Se ($T = 17.5$ sec) and cyclic activation analysis," *Radiochem. Radioanal. Letters*, **28**(5–6), 369–378.

Ettinger, K. V. (1988), "In vivo nuclear activation analysis" in R. Cesareo (ed.), *Nuclear Analytical Techniques in Medicine*, Elsevier, Amsterdam.

Fairweather-Tait, S. J., M. J. Minski, and D. P. Richardson (1983), "Ion absorption from a malted cocoa drink fortified with ferric orthophosphate using the stable isotope ^{58}Fe as an extrinsic label," *British Journal of Nutrition*, **50**, 51–60.

Franklin, D. M., D. R. Chettle, and C. Scott (1987), "Studies relating to the accuracy of in vivo measurements of liver and kidney cadmium," *J. Radioanal. Nucl. Chem.*, **114**(1), 155–163.

Gharib, A., H. Rahimi, H. Pyrovan, N.J. Raoffi, and H. Taherpoor (1985), "Study of trace elements in milk by nuclear analytical techniques," *J. Radioanal. Nucl. Chem.*, **89**(1), 31–44.

Girardi, F. and E. Sabbioni (1968), "Selective removal of radiosodium from neutron-activated materials by retention on hydrated antimony pentoxide," *J. Radioanal. Chem.*, **1**, 169–178.

Gladney, E. S., W. D. Moss, M. A. Gautier, and M. G. Bell (1989), "Determination of U in urine: comparison of ICP–mass spectrometry and delayed neutron assay," *Health Physics*, **57**(1), 171–175.

Gokmen, I. G., G.E. Gordon, and N.K. Aras (1987), "Application of different activation analysis techniques for determination of trace elements in human blood," *J. Radioanal. Nucl. Chem.*, **113**(2), 453–459.

Greenberg, R. R., R. F. Fleming, and R. Zeisler (1984), "High sensitivity neutron activation analysis of environmental and biological standard reference materials," *Environmental International*, **10**, 129–136.

Hancock, R.G.V., M.D. Grynpas, and B. Alpert (1987), "Are archaeological bones similar to modern bones? An INAA assessment," *J. Radioanal. Nucl. Chem.*, **110**(1), 283–291.

Heydorn, K. (1984), *Neutron Activation Analysis for Clinical Trace Element Research Vols I and II*, CRC Press, Boca-Raton, Florida.

Irigaray, J. L., H. Elmir, D. Pepin, and P.Y. Communal (1987), "Study of arsenic reabsorption in the body by the neutron activation analysis after thermal springs treatment," *J. Radioanal. Nucl. Chem.*, **113**(2), 469–476.

Ishii, T., K. Horiuchi, H. Nakahara, and Y. Murakami (1980), "Trace elements in the urine of diabetics determined by instrumental neutron activation analysis," *Radioisotopes*, **29**(6), 282–284.

Iyengar, G., W. E. Kollmer, and H. J. M. Bowen (1978), *Elemental Composition of Human Tissues and Body Fluids*, Springer Verlag, Berlin.

Kacperek, A., C.J. Evans, J. Dutton, W.D. Morgan, and A. Sivyer (1987), "A system for the determination of silicon in the human lung using neutrons from a 2 MV Van de Graaff generator," *J. Radioanal. Nucl. Chem.*, **114**(1), 165–172.

Kerr, S. A. and N. M. Spyrou (1978), "Fluorine analysis of bone and other biological materials: a cyclic activation method," *J. Radioanal. Chem.*, **44**(1), 159–173.

Kishi, T. (1987), "Forensic neutron activation analysis. The Japanese scene," *J. Radioanal. Nucl. Chem.*, **114**(2), 275–280.

Krishnan, S. S., M.T. Bayley, A.J.W. Hitchman, S.C. Lin, K.G. McNeill, and J.E. Harrison (1987), "Small sample in-vivo neutron activation analysis using californium sources," *J. Radioanal. Nucl. Chem.*, **114**(1), 173–180.

Kvicala, J. and J. Havelka (1988), "Frequency of concentrations of some trace elements in serum by INAA," *J. Radioanal. Nucl. Chem.*, **121**(2), 261–270.

Larsson, L., M. Alpsten, and S. Mattsson (1987), "In vivo analysis for nitrogen using a ^{252}Cf source," *J. Radioanal. Nucl. Chem.*, **114**(1), 181–185.

Mazzilli, B. and C. S. Munita (1986), "Mercury determination in dentist's hair and nails by instrumental neutron activation analysis," *Cienc. Cult.*, **38**(3), 522–526.

McDowell, L.S., P.R. Giffen, and A. Chatt (1987), "Determination of selenium in individual food items using the short-lived nuclide ^{77m}Se," *J. Radioanal. Nucl. Chem.*, **110**(2), 519–529.

Michel, R., F. Loer, M. Nolte, M. Reich, and J. Zilkens (1987), "Neutron activation analysis of human tissues, organs and body fluids to describe the interaction of orthopaedic implants made of cobalt–chromium alloy with the patients organisms," *J. Radioanal. Nucl. Chem.*, **113**(1), 83–95.

Noguchi, K., M. Shimizu, and E. Sairenji (1985), "Neutron activation analysis of mercury contents in head hair of dentists in Japan," *J. Radioanal. Nucl. Chem.*, **90**(1), 217–223.

Obrusnik, I., O. Skrivanek, M. Umlaufova, and V. Hovorka (1985), "Neutron activation analysis of neonate and maternal hair samples in areas with different levels of pollution," *J. Radioanal. Nucl. Chem.*, **89**(2), 561–570.

Ohta, Y., A. Nakano, M. Matsumoto, and M. Hoshi (1987), "Comparison of selenium content in human hair from different individuals in different countries by ^{76}Se(n,γ)^{77m}Se reaction," *J. Radioanal. Nucl. Chem.*, **114**(1), 75–82.

Papassotitiou, V., S. Georgala, J. Stratigos, N. Panayotakis, A. Hadjiantoniou, and A.A. Katsanos (1987), "Trace elements in skin epitheliomas," *J. Radioanal. Nucl. Chem.*, **109**(1), 89–100.

Parr, R. M. (1981), "Biomedical sciences," in S. Amiel (ed.), *Nondestructive Activation Analysis*, Elsevier, Amsterdam, pp. 139–174.

Parr, R. M. (1988), "On the role of neutron activation analysis in the certification of a new reference material for trace-element studies, mixed human diet, H-9," *J. Radioanal. Nucl. Chem.*, **123**(1), 259–271.

Prasad, A. S. (ed.) (1976) *Trace Elements in Human Health and Disease, Volumes 1 and 2*, Academic Press, New York.

Pritchard, J. G. and S. O. Saied (1986), "Studies on the determination of mercury in human beard shavings by neutron-activation and γ-ray analysis," *Analyst*, **111**, 29–35.

Stroube Jr, W. B., W.C. Cunningham, and G.J. Lutz (1987), "Analysis of foods for iodine by epithermal neutron activation analysis," *J. Radioanal. Nucl. Chem.*, **112**(2), 341–346.

Taskaev, E., M. Karaivanova, and T. Grigorov (1988), "Determination of platinum and gold in biological materials by neutron activation analysis," *J. Radioanal. Nucl. Chem.*, **120**(1), 75–82.

Tomza, U., T. Janicki, and S. Kossman (1983), "Instrumental neutron activation analysis of trace elements in hair: a study of occupational exposure to a nonferrous smelter," *Radiochem. Radioanal. Lett.*, **58**(4), 209–220.

Underwood, E. J. (1977), *Trace Elements in Human and Animal Nutrition*, Academic Press, New York, 4th Ed.

Valkovic, V. (1988), *Human Hair, Volumes I and II*, CRC Press, Boca Raton, Florida.

Versieck, J. and R. Cornelis (1989), *Trace Elements in Human Plasma or Serum*, CRC Press, Boca Raton, Florida.

Whitley, J. E., T. Stack, C. Miller, P. J. Aggett, and D. J. Lloyd (1987), "Determination of ^{58}Fe and ^{65}Cu enriched stable isotopic tracers in studies of mineral metabolism of babies," *J. Radioanal. Nucl. Chem.*, **113**(2), 527–538.

Yang, J. Y., S. M. Lin, C. H. Chiang, and M. H. Yang (1985), "Arsenic and mercury levels in head hair and fingernail as exposure indicators studied by instrumental neutron activation analysis," *Ho Tzu K'o Hseuh*, **22**(2), 97–103.

Zeisler, R. and R. R. Greenberg (1982), "Ultratrace determination of platinum in biological materials via neutron activation analysis and radiochemical separation," *J. Radioanal. Chem.*, **75**(1–2), 27–37.

Zeisler, R. and I. Young (1987), "The determination of chromium-50 in human blood and its utilization for blood volume measurements," *J. Radioanal. Nucl. Chem.*, **113**(1), 97–105.

Zhuk, L.I., A.A. Kist, I.N. Mikholskaya, N.S. Osinskaya, T. Tillayev, S.I. Tursunbayev, and S.V. Agzamova (1988), "Elemental blood composition of the inhabitants of Uzbekistan," *J. Radioanal. Nucl. Chem.*, **120**(2), 369–377.

ENVIRONMENTAL APPLICATIONS

Environmental chemistry covers a wide area of study concerned with air, water, plants, soil and animals (Bowen, 1979). The role of trace elements in living matter has been discussed in the previous chapter on biomedical applications. Soil will be considered under geochemical applications in the following chapter and environmental applications described in this chapter will be confined to the determination of trace elements in air, water and vegetation. Activation analysis is used as a routine technique in the study of environmental problems and there are two monographs describing applications to water, air, plants, soils and sediments (Das et al., 1983; Tolgyessy and Klehr, 1987). The matrices encountered can be quite different and include air particulates collected on cellulose paper filters, large-volume water samples, plants and soil. The matrix effects are not a serious problem for the activation analysis of cellulose filters or vegetation but, even if the collection medium itself does not interfere seriously with the analysis, the sample could contain high concentrations of elements such as iron, manganese and chromium, and the heavy metals, which do form interferences. In the case of water samples, or air pollutants collected on snow, the large volumes involved may create sample preparation problems.

AIR POLLUTANTS

There are several comprehensive reviews on the subject of analysis of atmospheric pollutants in general (Malissa, 1978) and activation analysis in particular (Alian and Sansoni, 1985; Dams et al., 1976; Dams, 1985; Ragaini, 1978). They describe the problems of sampling air pollution particulates, provide details of the impurities to be found in filter materials and review the applications of activation analysis to atmospheric pollutants. In particular Ragaini (1978) has described the different techniques of activation analysis used in characterization of atmospheric aerosols, including fast neutron activation and radiochemical separation techniques. Winchester and Desaedeleer (1981) describe some case histories where neutron activation analysis has been used to study the atmospheric environment.

Air samples are usually collected over a period of time on a filter medium such as cellulose paper, polystyrene, glass fiber or an organic membrane. The filter material will contain trace elements which may affect the analysis. In particular aluminum, barium, iron, magnesium and zinc are present in significant concentrations in many of the materials. Cellulose paper has the lowest concentration of these trace elements and it is therefore preferred for activation analysis. Cascade impactors are also used to collect air particulate and are usually made of polyethylene or polycarbonate.

The weights of samples collected for analysis can vary widely from a few milligrams to tens of grams. Interference effects in the gamma ray spectrum will be dependent on the air particulates themselves and will vary enormously. However there are no serious problems associated with the analytical technique and a routine scheme for the neutron activation analysis of air particulates (Dams et al., 1970), which is illustrated in Figure 14.1, provides results for up to 33 elements in an air filter sample. The samples and standards were irradiated for five minutes in a thermal neutron flux of 2×10^{16} n m^{-2} s^{-1} and counted for 400 s after a 3 min decay period, for Al, V, Ti, S and Ca. After a 15 min decay the samples were recounted for 1000 s for Mg, Br, Cl, I, Mn, Na and In. A second longer irradiation of 2–5 h in the higher flux of 1.5×10^{17} n m^{-2} s^{-1} was used to measure the long-lived radionuclides. After a 20–30 h decay period they were recounted for 2000 s for K, Zn, Br, As, Ga, Sb, La, Sm, Eu, W and Au. After a further decay of 20–30 d Sc, Cr, Fe, Co, Ni, Zn, Se, Ag, Sb, Ce, Hg and Th were measured. Typical values for suspended particulate from East Chicago are shown in Figure 14.1 with the detection limits for those samples. In this particular case the samples were collected on polystyrene filters which contained substantial amounts of chlorine so the chlorine blank was too high to provide an accurate value for the sample.

Another typical scheme which is used routinely for atmospheric samples on 100 mm Whatman 41 paper involves two irradiations in a thermal neutron flux of 1.5×10^{17} n m^{-2} s^{-1} (Kronborg and Steinnes, 1975). The sample is irradiated for 5 min and counted for 5 min after a decay period of 8 min for Na, Cl, Al, V, Br and Mn. A longer irradiation of 2 d is then followed by a series of three counts after 3–5 d, 5–7 d and >14 d for 10, 20 and 30 min respectively. As, La, Sb, Fe, Co, Cr, Zn, Ag, Sc, Se and Sb are determined during the course of the three counts.

An interesting way of monitoring airborne pollutants is with the collection of snow (Zikovsky and Badillo, 1987). The snow was collected

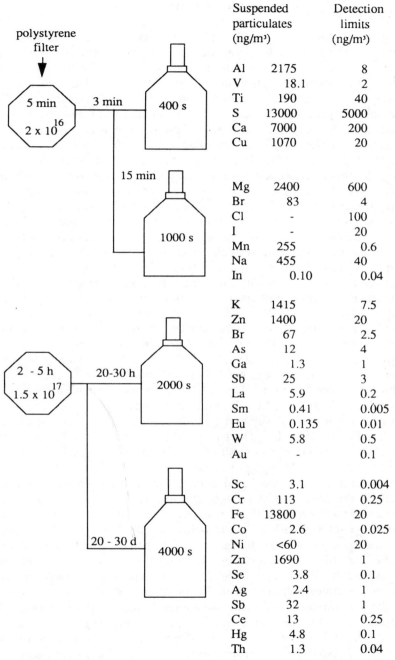

Figure 14.1. Neutron activation analysis of air pollution particulates. (Adapted with permission from Dams et al., 1970.)

183

using plastic tools from roads and on the roof of a tall building. The samples were allowed to thaw, the solid fraction filtered off and the liquid portion was dried. The dried fraction was irradiated for 1, 10 and 100 min in a thermal neutron flux of 10^{16} n m^{-2} s^{-1}. The samples were counted after decay times which were between 20 s and 16 d. The insoluble fraction, on filter paper, was irradiated for 1 and 100 min and counted at an interval of 20 days minimum between counts. In total 23 elements were detected in the samples: Al, As, Br, Ce, Cl, Co, Cr, Cs, Cu, Fe, Hg, K, La, Mg, Mn, Mo, Na, Sb, Sc, Sr, Ti, V and Zn.

Epithermal neutron activation is rarely used to improve the detection limits of trace elements in environmental samples because many of the interfering heavy metals are not affected by a thermal neutron filter. However one example is the work of Wyttenbach et al. (1987) who studied the aerosols adhering to spruce needles and pine needles from a Swiss landscape by dissolving the wax and collecting the solids. The samples of solid removed from the twigs, weighing about 10 mg, were analyzed for 33 elements with three thermal neutron irradiations and one epithermal neutron irradiation. A 1 min irradiation in an epithermal neutron flux of 1.9×10^{15} n m^{-2} s^{-1} was used to measure bromine, chlorine and iodine. A normal scheme of irradiations in a thermal neutron flux of 2.5×10^{17} n m^{-2} s^{-1} for short-, medium- and long-lived radionuclides were as follows: 20 s irradiation, 5 min decay and 6 min count for Al, Ca, Cl, Mg, Na, Ti and V, and after a 60 min decay a count for 10 min for Ba, Cl, Mn and Na; a 20 min irradiation, 30 h decay and 2 h count for As, Au, Br, Ga, K, La, Na, Sb, W and Zn; an 80 min irradiation, 10 d decay and 7 h count for Au, Ba, Br, Ce, Co, Cr, Cs, Eu, Fe, Hf, Hg, La, Th, Rb, Sb, Sc, Se, Sm, Ta and Zn. In this way all 33 elements were measured in concentrations ranging from 0.22 μg kg^{-1} of gold to 24 mg kg^{-1} of aluminum.

Fast neutrons have been used to determine percentage levels of heavy metals in the environment around welders (Tolgyessy et al., 1987). The welding aerosol was collected on a membrane filter and irradiated in a neutron generator with a neutron flux of about 10^9 n s^{-1} for 300 s followed by a 300 s count to determine Cr, Mn, Mo and then for 360 s for Fe and Mo. A rather more comprehensive list of elements was determined by Bahal and Pepelnik (1986) who used a high intensity neutron generator with a neutron flux of about 3×10^{12} n s^{-1}. A scheme for the determination of some 34 elements was used to measure the composition of Milanese air dust. A sample weighing 47 mg was irradiated for 30 min and counted after 45 s, 14 min and 2.5 h for 10 min, 1.4 h and 1.8 h, respectively. Cyclic activation was also used for short-lived radionuclides, and for very long-lived radionuclides the sample was recounted for 64 h after at least 6 h decay. This provided the maximum information and sensitivity with detection limits between 10 and 200 mg kg^{-1} for many of the trace elements.

Photon activation is generally less sensitive than neutron activation but it can be used to advantage for titanium, nickel, arsenic, iodine and lead which are difficult to measure with neutron activation (Aras et al., 1973). Fourteen elements, Na, Cl, Ca, Ti, Cr, Ni, Zn, As, Br, Zr, Sb, I, Ce and Pb were determined routinely using an irradiation with bremsstrahlung from electrons of 35 MeV from the NBS electron linac and measurement of gamma rays. The sample was pelletized and irradiated for several hours. A scheme of four counts was used with decay periods of 1–2 h for Cl, Ni and Pb; 1–3 d for Ca, Ni, Zn, As, Sb and Pb; 10–12 d for Ca, Ti, Zn, As, Br, Zr and Pb; and >3 weeks for Ti, Cr, I and Ce. The detection limits were between 0.1 and 1 ng m^{-3} for Cl, Ti, Cr.

The ash from coal fired power stations is of interest to environmentalists. Coal fly ash is readily activated and is not easy to analyze. The weight of sample must be kept low to reduce the activity induced in elements such as sodium, cobalt and scandium. High aluminum content makes a short irradiation difficult. A scheme for the determination of 23 elements in coal fly ash used neutron activation with both prompt and delayed gamma ray analysis (Vogt and Schlegel, 1985). Samples weighing 100 mg were first irradiated in a neutron flux of 1.3×10^{17} n m^{-2} s^{-1} for 60 s and left to decay for 15 min before being counted for 300 s on a Ge(Li) detector for Na, Al, Ca, Ti, Mn and V. After a 2 week decay the samples were re-irradiated for 1 h in a flux of 5×10^{17} n m^{-2} s^{-1}. After a decay of 3–5 d the samples were counted for K, As, La, U, Mo and Sm. They were recounted after 10–13 d for the long-lived radionuclides of Ba, Fe, Sb, Cs, Co, Cr, Th, Sc, Ce, Eu and Hf. Finally prompt gamma ray analysis was used to determine B, Al, Si, K, Ca, Ti, Fe, Cd, Sm and Gd in a neutron beam of 5×10^{8} n s^{-1}. Results were obtained for 23 elements in NBS coal fly ash 1633A. In particular the gadolinium value of 16.3 mg kg^{-1} and the cadmium value of 1.01 mg kg^{-1} showed that prompt gamma ray analysis provides useful data for environmental samples.

WATER

Liquid samples are generally more difficult to handle and create additional problems for activation analysis, which do not affect other techniques. Representative water samples are normally fairly large in volume and some irradiation sites are not built to accommodate large liquid samples. Most other analytical techniques require the sample to be in solution and therefore water presents an ideal sample free from preparation problems. Sampling for activation analysis usually requires a preconcentration step to ensure adequate sensitivity. If the solid insoluble component of the water is to be measured, the water will be filtered and the particulates analyzed. If the dissolved material is to be measured then the water can be irradiated as a liquid or freeze-dried to remove the liquid and irradiated as a solid. Alternatively the water may be passed through an ion exchange column to extract the elements of interest.

Rainwater has been analyzed for 35 elements using instrumental neutron activation (Schutyser et al., 1978). First the rainwater was filtered and the solid portion dried and pelletized with the filter support. The liquid fraction was freeze dried in 250 cm^3 aliquots and pelletized in clean plastic bags. The short irradiation of 5 min at 2.6×10^{16} n m^{-2} s^{-1} was followed by two counting periods to measure Na, Mg, Al, Cl, Ca, Ti, V, Mn, Cu, Br, Sr, In and I. The long irradiation of 32 h at a flux of 1.6×10^{16} n m^{-2} s^{-1} was followed by a count after 1 d for Na, K and W while As, Br, Sb, La, Sm and Au were measured after 5 d. Finally Sc, Cr, Fe, Co, Zn, Se, Ag, Cs, Ce, Eu, Lu and Th were determined after 20 d. Rainwater was analyzed as part of an urban monitoring program for bromine, chlorine and iodine (Landsberger et al., 1988). The simultaneous instrumental neutron activation analysis gave detection limits of 2, 30 and 1 μg kg^{-1} for bromine, chlorine and iodine, respectively.

River water is a simple material, which when concentrated can be readily analyzed for trace elements. Capannesi et al. (1984) measured 46 elements in the suspended particulate matter and liquid fraction of river water samples. The irradiation scheme is illustrated in Figure 14.2. Solid portion samples weighing 20–100 mg (or liquid fractions weighing 2 g) were irradiated for 10–30 s (or 60–300 s) in a thermal neutron flux of 1.1×10^{17} n m^{-2} s^{-1} in a pneumatic irradiation system. They were counted after a 5–10 min decay (2–3 min for liquids) and again after 20–30 min. A longer irradiation was made for 30 h in a flux of 2.6×10^{16} n m^{-2} s^{-1} using the same weight of solid fraction or 20 g of liquid fraction. After a decay period of 3–5 d the first count was for 1–3 h and after 40 d the second count was for 10–20 h. Finally another measurement was made after 200 d decay. Clearly the full scheme took a long time to complete. Typical results for the liquid fractions are shown in Figure 14.2 together with the detection limits for the 46 elements measured.

Preconcentration on an ion exchange resin is commonly used to collect specific elements. An example of such a procedure is the preconcentration of gold on activated charcoal (Hamilton and Ellis, 1983). It is a method used for hydrogeochemical prospecting by measuring the gold content of surface water to indicate deposits. The water samples were filtered and

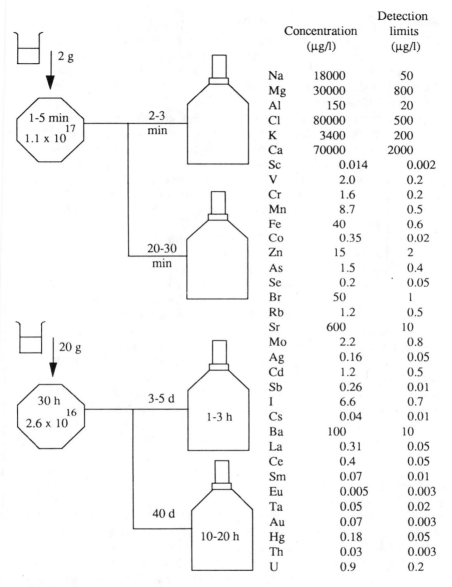

	Concentration (μg/l)	Detection limits (μg/l)
Na	18000	50
Mg	30000	800
Al	150	20
Cl	80000	500
K	3400	200
Ca	70000	2000
Sc	0.014	0.002
V	2.0	0.2
Cr	1.6	0.2
Mn	8.7	0.5
Fe	40	0.6
Co	0.35	0.02
Zn	15	2
As	1.5	0.4
Se	0.2	0.05
Br	50	1
Rb	1.2	0.5
Sr	600	10
Mo	2.2	0.8
Ag	0.16	0.05
Cd	1.2	0.5
Sb	0.26	0.01
I	6.6	0.7
Cs	0.04	0.01
Ba	100	10
La	0.31	0.05
Ce	0.4	0.05
Sm	0.07	0.01
Eu	0.005	0.003
Ta	0.05	0.02
Au	0.07	0.003
Hg	0.18	0.05
Th	0.03	0.003
U	0.9	0.2

Figure 14.2. Neutron activation analysis of river water. (Adapted with permission from Capannesi et al., 1984.)

the pH of the filtrate adjusted to between 3 and 4. Activated charcoal weighing 0.1 g was added to the liquid and it was shaken for 5 min. The charcoal was then filtered off and dried before it was irradiated for 24 h in a thermal neutron flux of 5×10^{16} n m^{-2} s^{-1}. After a 4 d decay the irradiated charcoal sample was counted for gold. The detection limit was 0.3 μg m^{-3} total gold.

Seawater presents a particularly difficult matrix for neutron activation analysis since activity of sodium and chlorine induced in the matrix makes analysis of short-lived radionuclides difficult. The fish and shellfish living in water are good indicators of the quality of the water and so they are studied for toxic elements such as the heavy metals. Simsons and Landsberger (1987) analyzed marine biological material for 24 trace elements. The irradiation scheme is shown in Figure 14.3. The short-lived gamma ray spectrum of lobster hepatopancreas is dominated by the lines from sodium and chlorine after irradiation of only 10 s in a thermal neutron flux of 5×10^{16} n m^{-2} s^{-1}. Following a 150 s delay the samples were counted for 10 min for Ca, Cu, Cl, Mg, Na and V. After 2.5 h, potassium and manganese were measured for 1000 s. Epithermal activation was preferred for Sr, I and Br and an irradiation of 3 min in a flux of 2×10^{15} n m^{-2} s^{-1} was followed by a 45 min decay and a 30 min count. The medium half-life radionuclides of Sb, As, La and Sm were measured after a 1 h irradiation in the thermal neutron flux and counted for 1 h after a 1 week decay. Long-lived Ag, Ce, Cr, Co, Eu, Fe, Hf, Rb, Sc, Se and Zn were measured after a 2 week decay. A long epithermal neutron irradiation was used to measure nickel via the ^{58}Ni(n,p)^{58}Co reaction. Uranium was measured with delayed neutron counting while cadmium and sulphur were determined using prompt gamma ray analysis of a 0.5 g sample in a flux of 7×10^{12} n m^{-2} s^{-1}. Figure 14.3 gives typical values and detection limits.

Zeisler et al. (1988) have analyzed marine bivalve samples for 33 elements by instrumental neutron activation. Prompt gamma ray analysis is used to determine H, B, C, N, Na, P, S, Cl, K and Cd. A short irradiation is used for Na, Mg, Al, Cl, K, Ca, V, Mn, and I. A long irradiation is made for Sc, Cr, Fe, Co, Zn, As, Se, Br, Rb, Sr, Mo, Ag, Cd, Sb, Cs, Ba, La, Ce, Sm, Eu, Hf, Au, Hg and Th. Further elements are measured using X-ray fluorescence and tin is determined with a radiochemical separation prior to measurement. Yagi and Masumoto (1987) used proton activation to analyze oyster tissue and mussel. One gram samples were dissolved in acid and dried before being mixed with silica gel. A proton energy of 13 MeV and a beam current of 2–4 μA was used for 2 hours. After a 1 d decay the samples were counted for Ca, Zn, Zr, Mo, and 1 month later they were recounted for Ti, V, Fe,

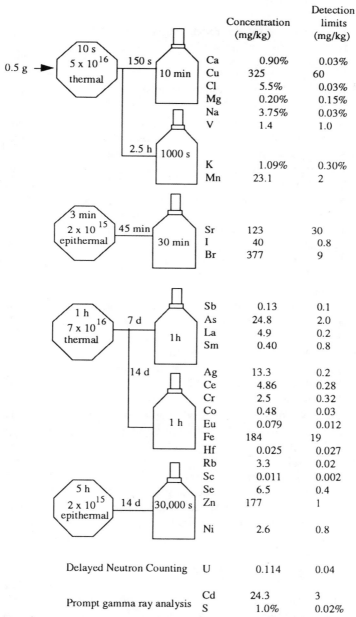

		Concentration (mg/kg)	Detection limits (mg/kg)
Ca		0.90%	0.03%
Cu		325	60
Cl		5.5%	0.03%
Mg		0.20%	0.15%
Na		3.75%	0.03%
V		1.4	1.0
K		1.09%	0.30%
Mn		23.1	2
Sr		123	30
I		40	0.8
Br		377	9
Sb		0.13	0.1
As		24.8	2.0
La		4.9	0.2
Sm		0.40	0.8
Ag		13.3	0.2
Ce		4.86	0.28
Cr		2.5	0.32
Co		0.48	0.03
Eu		0.079	0.012
Fe		184	19
Hf		0.025	0.027
Rb		3.3	0.02
Sc		0.011	0.002
Se		6.5	0.4
Zn		177	1
Ni		2.6	0.8
Delayed Neutron Counting	U	0.114	0.04
Prompt gamma ray analysis	Cd	24.3	3
	S	1.0%	0.02%

Figure 14.3. Neutron activation analysis of marine samples. (Adapted with permission from Simsons and Landsberger, 1987.)

189

As and Sr. The detection limits (in mg kg^{-1}) were as follows: Mo 0.1; Zr 0.2; Ti 0.3; V, As and Sr 0.5; Fe 0.7; Zn 5; and Ca 25.

VEGETATION

Vegetation may be used in several ways to indicate the presence and locate the sources of pollution. Plants themselves are sometimes very good indicators of what elements are present in the soil and much work is done on plants to study the mechanisms of the uptake of trace elements from soil. Information on the trace element composition of soils and plants and on the mechanisms controlling the trace element cycle is given in a publication by Kabata-Pendias and Pendias (1984). Leaves can be analyzed by neutron activation to measure trace elements which are taken up by the plant system and also to measure airborne pollutants deposited on them. Vegetation has even been used in mineral exploration studies to locate deposits of gold. Leaves are frequently analyzed because of the ease with which they may be sampled. In most cases the samples are analyzed after the removal of any surface contamination, although in some pollution studies it is the surface contamination which is measured. The leaves may be edible or used as tobacco so the trace elements are important in the area of human health.

Tea leaves are important in human health, in view of the wide consumption of tea. Ahmad et al. (1983) made a study of the trace elements in several different types of dried tea leaves using instrumental neutron activation analysis. Toxic elements such as As, Br, Sb, Hg and Se were of greatest interest. Samples weighing 2 g were irradiated for 2 min, 1 h, or 48 h in a flux of 2×10^{17} n m^{-2} s^{-1}. The samples irradiated · for 2 min were allowed to decay for 2 h before they were counted for Na, K and Mn. After the 1 h count the samples decayed for 2 d before counting for As, Br and La. The long irradiation was followed by a 1 week decay when Ca was measured, followed by a further 1 week when Sc, Cr and Fe were counted and finally after 4 weeks Co, Zn, Se, Rb, Sb, Cs, Eu, Lu and Hg were measured. In total 19 elements were determined in the tea leaves.

Large samples of leaves may be collected for analysis. In general the leaves are dried and crushed so that a small amount can be taken as a representative sample. The amount of material is variable but generally below 1g. For example Capannesi et al. (1988) collected 500 g of leaves, wet weight, from an evergreen oak to provide a single sample as an indicator in the study of pollution. Samples were blown with nitrogen gas to clean dust and particles from the surface and then the leaves were dried. Subsequently the dried leaves were pulverized and the homogeneity tested for 1 g samples. Samples weighing 1 g were irradiated for 6 s in a flux of 1.3×10^{17} n m^{-2} s^{-1} in a TRIGA® Mark II reactor. A 10 g sample was irradiated for 30 h in a flux of 2.6×10^{16} n m^{-2} s^{-1}, in a rotating rack for uniform flux distribution. The short irradiation samples were counted after 7 min for Mg, Al, Ti, V and Cu; after 25 min for Cl and I; and after 8–12 h for Na, K and Mn. The long irradiation samples were counted after 5–8 d for Ca, As, Br, Mo, Cd, La, Sm, W, Au and U; and after 30–50 d for Sc, Cr, Fe, Co, Ni, Zn, Se, Rb, Sr, Ag, Sn, Sb, Ba, Ce, Nd, Eu, Tb, Yb, Hf, Ta, Hg and Th. Photon activation of a 10 g sample was used to measure Pb and As.

Samudralwar et al. (1987) analyzed edible leaves such as spinach, cabbage and coriander leaves. Samples were hand picked, washed, dried and powdered before analysis in small sample weights, 20 to 40 mg. The irradiation scheme consisted of four irradiations for 5 min, 1–2 h, 5 h, and 10–15 h. The short irradiation was used to measure Na, K, Cl, Mn, Mg, Cu and Ga after a 1 h decay. After the 1–2 h irradiation the sample decayed for 1 d, then it was counted for Na, K, Cu, Ga, Br, La, Hg, Cd, Mo, U and Eu. The 5 h irradiation followed by a 2 d decay was used to determine Br, La, Hg, Fe, Ce, Cd, Mo and Eu; and finally the long irradiation and decay of 5 d was used to measure Hg, Mo, Ce, Cd, Sc, Fe, Zn, Co and Cr. Some elements were measured more than once during the series of measurements.

The US Food and Drug Administration has a routine procedure for the neutron activation analysis of food including leafy vegetables (Cunningham and Stroube Jr, 1987). The procedure, shown in Figure 14.4, consists of only two irradiations to measure the short and long-lived radionuclides. The dried sample, weighing about 300 mg, is irradiated for 15 s in a thermal neutron flux of 4.9×10^{17} n m^{-2} s^{-1}. Following a 2 min decay time the samples are counted for 10 min. The second irradiation is for 4 h and the samples are counted after decay periods between 5 d and 3 weeks. Nineteen elements are measured in leafy vegetables and typical results for Ag, Br, Ca, Cl, Co, Cr, Cs, Cu, Fe, K, Mg, Mn, Na, Rb, Sb, Sc, Se, V and Zn are shown in Figure 14.4. The concentrations of some of the elements, including Cu, Sb and Se, are below the detection limits.

Trace elements in the leaves of tobacco are of significance to human health. Thermal neutron activation has been used to determine 28 elements in tobacco and cigarette paper (Iskander, 1985). Samples were irradiated for 3 min and for 8 h in a flux of 2×10^{16} n m^{-2} s^{-1}. Samples were measured 1 min after the short irradiation for Al, V, Ti, Ca and Mg, and after a 30 min delay for Cl and Mn. The samples irradiated for

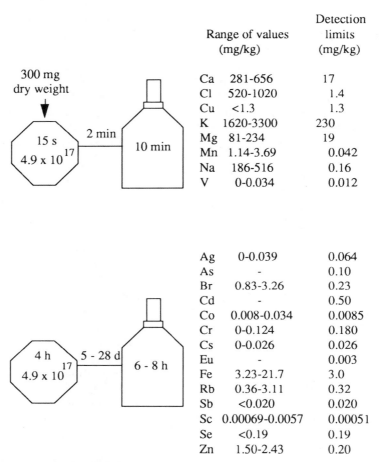

Figure 14.4. Neutron activation analysis of leafy vegetables. (Adapted with permission from Cunningham and Stroube Jr, 1987.)

8 h were allowed to decay for 12 h before they were counted for 4000 s to measure K, Na, As, Br and La. They were recounted for 11 h after a 21 d decay for Ba, Rb, Th, Cr, Ce, Hf, Fe, Sb, Sr, Ni, Sc, Se, Zn, Cs, Co and Eu. Positive values were obtained for all 28 elements in tobacco.

Caplan et al. (1987) took between 100 and 330 mg of grasses, heathers and herbs to study the uptake of metals from soils of different acidities. The elements studied included Na, Sc, Cr, Fe, Co, Zn, Br, Rb, Ce, Ba and La. They used a 24 h irradiation in a flux of 5×10^{16} n m^{-2} s^{-1} and counted after 3 d, 7 d and 3 weeks. They concluded that Na, K, Ca, Fe, Zn, Rb, Ba were better measured with atomic absorption and inductively coupled plasma emission spectrometry but that Sc, Cr, Co, Br, Cs and La could be satisfactorily measured by neutron activation analysis.

Neutron activation was used to determine as many elements as possible in peat samples from ombrotrophic bogs by Njastad et al. (1987). The material was first air dried and disintegrated. Then samples with a volume of about 2.5 cm^3, weighing between 0.1 and 1.0 g, were irradiated in a flux of 5×10^{16} n m^{-2} s^{-1} for 24 h. After a 3 d decay the samples were counted for 5 min for Na, K, As, Br, Cd and La. After a further 10 d the samples were measured for 20 min for Ca, Mo, Sb, Sm and Au. Finally after 20 d they were counted for 60 min for Sc, Cr, Fe, Co, Zn, Se, Rb, Sr, Cs, Ba, Ce, Eu, Th. This provided results for a total of 24 elements from one irradiation, within 3 weeks of irradiation.

Some of the trace elements which are important in plant growth are not easily determined by neutron activation. Therefore Le Roux et al. (1988) used proton-induced X-ray emission analysis to determine Mg, Al, Si, P, S, Cl, K and Ca in dried roots, to study the toxic effects of aluminum in vines. Powdered samples were pelletized and irradiated with 1 MeV protons from a Van de Graaff generator. The beam current was about 2 nA and the irradiations lasted for between 8 and 15 min. A 14 MeV neutron generator with a flux of 2×10^{11} n s^{-1} was used to measure N, P, K and Si in fertilizers and vegetables. Typical results showed fertilizers containing 30% N, 12% P, 40% K and 0.5% Si. Peas which were analyzed contained 3% N, 0.4% P and 1% K. An automated pneumatic system capable of handling up to 40 samples per day, has been developed for the routine analysis of fertilizers and plants using 14 MeV neutrons (Kafala et al., 1986). Samples are irradiated for 1 min, allowed to decay for 5 min and counted for 5 min. Fertilizer, peas or wheat, are analyzed routinely for percentage levels of nitrogen, phosphorus, potassium and silicon.

REFERENCES

Ahmad, S., M. S. Chaudhary, A. Mannan, and I. H. Qureshi (1983), "Determination of toxic elements in tea leaves in instrumental neutron activation analysis," *J. Radioanal. Chem.*, **78**(2), 375–383.

Alian, A. and B. Sansoni (1985), "A review on activation analysis of air particulate matter," *J. Radioanal. Nucl. Chem.*, **89**(1), 191–275.

Aras, N.K., W.H. Zoller, and G.E. Gordon (1973), "Instrumental photon activation analysis of atmospheric particulate material," *Anal. Chem.*, **45**(8), 1481–1490.

Bahal, B.M. and R. Pepelnik (1986), "Multielement analysis of a Milanese air-dust sample by 14 MeV neutron activation," *J. Radioanal. Nucl. Chem.*, **97**(2), 359–372.

Bowen, H. J. M. (1979), *Environmental Chemistry of the Elements*, Academic Press, London.

Capannesi, G., A. Cecchi, and P. A. Mando (1984), "Trace elements in suspended particulate and liquid fraction of the Arno river waters," in *Proceedings of the Fifth International Conference on Nuclear Methods in Environmental and Energy Research*, University of Missouri, Missouri, CONF-840408, pp. 311–321.

Capannesi, G., S. Caroli, and A. Rosada (1988), "Evergreen oak leaves as natural monitor in environmental pollution," *J. Radioanal. Nucl. Chem.*, **123**(2), 713–729.

Caplan, J., E. Lobersli, R. Naeumann, and E. Steinnes (1987), "A neutron activation study of trace element content in plants growing on soils of different acidity," *J. Radioanal. Nucl. Chem.*, **114**(1), 13–19.

Cunningham, W. C. and W. B. Stroube Jr. (1987), "Application of an instrumental neutron activation analysis procedure to analysis of food," *The Science of the Total Environment*, **63**, 29–43.

Dams, R. (1985), "Application of multi-elemental neutron activation analysis in environmental research," in B. Sansoni (ed.), *Instrumentelle Multielement-analyse*, VCH, Weinheim, pp. 577–593.

Dams, R., J. A. Robbins, K. A. Rahn, and J. W. Winchester (1970), "Nondestructive neutron activation analysis of air pollution particulates," *Anal. Chem.*, **42**(8), 861–868.

Dams, R., F. De Corte, J. Hertogen, J. Hoste, W. Maenhaut and F. Adams (1976), "Activation Analysis – Part 1," in T. S. West (ed.), *Physical Chemistry Series Two, International Review of Science*, Butterworths, London, pp. 24–38.

Das, H. A., A. Faanhof, and H. A. van der Sloot (1983), *Environmental Radioanalysis*, Elsevier, Amsterdam.

Hamilton, T. W., and J. Ellis (1983), "Determination of gold in natural waters by neutron activation – γ-spectrometry after preconcentration on activated charcoal," *Anal. Chim. Acta,* **148**, 225–235.

Iskander, F. Y. (1985), "Neutron activation analysis of an Egyptian cigarette and its ash," *J. Radioanal. Nucl. Chem.,* **89**(2), 511–518.

Kabata-Pendias, A. and H. K. Pendias (1984), *Trace Elements in Soils and Plants*, CRC Press, Wolfe Medical Publications, London.

Kafala, S.I., B.V. Anisimov, and I.V. Tolkachev (1986), "Automatic activation analysis of fertilizers and plant samples by fast neutrons," *J. Radioanal. Nucl. Chem.,* **97**(2), 341–345.

Kronborg, O. J. and E. Steinnes (1975), "Simultaneous determination of arsenic and selenium in soil by neutron activation analysis," *Analyst*, **100**, 835–837.

Landsberger, S., J. J. Drake, and S. J. Vermette (1988), "Enriched concentrations of bromine, chlorine and iodine in urban rainfall as determined by instrumental neutron activation analysis," *Chemosphere*, **17**(2), 299–307.

Le Roux, E. G., M. Peisach, C. A. Pineda, and M. A. B. Pougnet (1988), "The toxic effect of aluminium in vines," *J. Radioanal. Nucl. Chem.*, **120**(1), 97–104.

Malissa, H. (ed.) (1978), *Analysis of Airborne Particles by Physical Methods*, CRC Press, Florida.

Njastad, O., R. Naeumann, and E. Steinnes (1987), "Variations in atmospheric trace element deposition studied by INAA of peat cores from ombrotrophic bogs," *J. Radioanal. Nucl. Chem.*, **114**(1), 69–74.

Ragaini, R. C. (1978), "Characterization of atmospheric aerosols by neutron activation analysis," in H. Malissa (ed.), *Analysis of Airborne Particles by Physical Methods*, CRC Press, Florida, pp. 93–123.

Samudralwar, D. L., H. K. Wankhade, and A. N. Garg (1987), "Multielemental analysis of IAEA intercomparison standard hay powder, V-10 and some edible plant leaves by neutron activation," *J. Radioanal. Nucl. Chem.*, **116**(2), 307–315.

Schutyser, P., W. Maenhaut, and R. Dams (1978), "Instrumental neutron activation analysis of dry atmospheric fall-out and rainwater," *Anal. Chim. Acta*, **100**, 75–85.

Simsons, A. and S. Landsberger (1987), "Analysis of marine biological certified reference material by various non-destructive neutron activation methods," *J. Radioanal. Nucl. Chem.*, **110**(2), 555–564.

Tolgyessy, J. and E. H. Klehr (1987), *Nuclear Environmental Chemical Analysis*, Ellis Horwood, Chichester.

Tolgyessy, J., L. Olah, and E.H. Klehr (1987), "Contribution to the radioanalytical determination of heavy metals in the working environment of welders," *J. Radioanal. Nucl. Chem.*, **114**(1), 101–104.

Vogt, J. R. and S.C. Schlegel (1985), "Elemental determinations in NBS 1633A fly ash standard reference material using INAA and PGNAA," *J. Radioanal. Nucl. Chem.*, **88**(2), 379–387.

Winchester, J. W. and G. G. Desaedeleer (1981), "Applications of trace element analysis to studies of the atmospheric environment," in S. Amiel (ed.), *Nondestructive Activation Analysis*, Elsevier, Amsterdam, pp. 187–236.

Wyttenbach, A., S. Bajo, and L. Tobler (1987), "Aerosols deposited on spruce needles," *J. Radioanal. Nucl. Chem.*, **114**(1), 137–145.

Yagi, M. and K. Masumoto (1987), "Instrumental charged-particle activation analysis of several selected elements in biological materials using the internal standard method," *J. Radioanal. Nucl. Chem.*, **111**(2), 359–369.

Zeisler, R., S. F. Stone, and R. W. Sanders (1988), "Sequential determination of biological and pollutant elements in marine bivalves," *Anal. Chem.*, **60**, 2760–2765.

Zikovsky, L. and M. Badillo (1987), "An indirect study of air pollution by neutron activation analysis of snow," *J. Radioanal. Nucl. Chem.*, **114**(1), 147–153.

CHAPTER

15

GEOLOGICAL APPLICATIONS

Geological material represents the major sample type to be analyzed by activation analysis techniques. Large numbers of soils or sediment samples are analyzed for up to 40 elements as part of geochemical mapping projects or exploration studies. The rare earth elements, which can be determined by neutron activation analysis, are of special interest in the study of the geochemistry of igneous rocks. Specific methods are used in the case of single elements for exploration studies such as delayed neutron counting for uranium and epithermal neutron activation for gold. Neutron activation with a preconcentration stage provides one of the most sensitive techniques for the determination of the platinum group elements. Borehole logging with a portable neutron source is used for *in situ* analysis of possible deposits of elements such as copper or nickel. A general book on radioanalysis in geochemistry (Das et al., 1989) describes the use of activation techniques; applications of neutron activation analysis in geochemistry and cosmochemistry are reviewed by Cornelis et al. (1976).

SOILS

Many countries have Geological Surveys which carry out complete analyses on soils and stream sediments requiring a number of analytical techniques. A book by Smith (1983) on the subject of soil analysis covers the instrumental techniques used and related procedures. Neutron activation analysis provides a multielemental facility for routine analysis of samples and has been widely used for this purpose. Salmon and Cawse (1983) review the use of instrumental neutron activation for soil and give examples of applications in total element analysis, the analysis of soil fractions, plant and animal nutrition, environmental pollution, forensic science and archaeology.

Soils contain percentage levels of aluminum which activates via the (n, γ) reaction to ^{28}Al. This gives rise to a high background activity which results in high count rates in the detector system and limits the sample size that can be irradiated. The small sample size is not a problem with soils which are usually homogeneous. Commonly the sample will be irradiated for a few minutes and left for a few minutes to allow the ^{28}Al to decay a little before counting. The determination of aluminum itself is difficult in the case of silicate rocks because of the fast neutron induced reaction ^{28}Si(n,p)^{28}Al. The interference effect can be quite significant, depending on the aluminum to silicon ratio in the sample, and the thermal to fast neutron flux ratio in the reactor. A similar reaction of fast neutrons on aluminum can cause an interference in the determination of magnesium: ^{27}Al(n,p)^{27}Mg. Elements such as aluminum and magnesium are better measured by other techniques.

A typical example of the procedures used for routine analysis of geological survey samples is that of Robotham et al. (1987). They analyzed about 100 samples for 23 elements as part of a study of elemental concentrations throughout Jamaica for geochemical maps. The irradiation scheme is shown in Figure 15.1. The original sample weighing 4 kg was subsampled at 500 g for laboratory analysis. Samples weighing 250 mg were irradiated in a thermal neutron flux of 2.5×10^{15} n m^{-2} s^{-1} for 4 min. The sample was counted twice, once after a 12 min decay for 5 min to measure V, Ti, Al and then after a 30 min decay (to allow the aluminum activity to decay away) for 10 min for Mn, Mg, K, Eu, Sm, Dy, Ca and Ba. The long irradiation was for 1 h followed by a 4 d delay to allow the sodium activity to decay before counting for the radionuclides of As, U, La, Lu and Sb which have relatively short half-lives. Radionuclides with similar half-lives to ^{24}Na, such as Cu, Pr, and Er are not usually detected except where they are present in high concentrations. After a 21 d decay the samples are recounted for Co, Sc, Th, Fe, Hf and Cr. At that time the main activity producing background will be due to ^{46}Sc and ^{60}Co. The calculated detection limits for soils are given in Figure 15.1.

Routine procedures used in Finland follow a similar scheme of five measurements for the determination of 41 elements (Rosenberg et al., 1983). The first short irradiation for 30 s is followed by a 2 min decay before counting for 1–5 min for Ti, Al, Ca and V. A 5–20 min count, after a 5 min irradiation and a 30 min decay period, is used to measure Mn and Dy. The intermediate radionuclides of Na and K are measured by irradiating for 20–60 min and waiting for 24 h before counting for 5–20 min. The long-lived radionuclides are determined after a 7–30 h irradiation, followed by a 4–7 d decay and 0.5–2 h count for As, Br, Mo, Sb, La, Sm, Lu, W, Au and U. Finally after a 4–6 week decay the samples are recounted for 0.5–4 h for Sc, Cr, Fe, Ni, Zn, Se, Co, Rb, Ag, Sn, Zr, Cs, Ba, Ce, Nd, Eu, Gd, Tb, Tm, Yb, Hf, Ta and Th.

A very similar procedure was used by Adepetu et al. (1988) for the

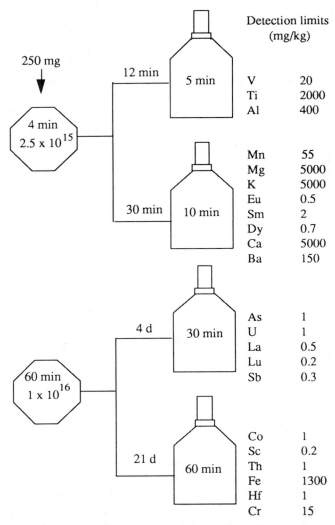

Figure 15.1. Neutron activation analysis of Jamaican soil. (Adapted with permission from Robotham et al., 1987.)

analysis of Nigerian Harmattan dust. Two irradiations were made, a short one for 3 min and a long one for 8 h, both in a thermal neutron flux of 2×10^{16} n m^{-2} s^{-1}. After the short irradiation the samples were counted twice, first after 1–2 min for 3 min to measure Al, Mg, Ti and V, and then after 30–45 min for Mn and Cl. The samples were counted for intermediate half-life radionuclides As, Au, Br, Ga, K, La, Na and Sm, 24–36 h after the long irradiation for 1 h. Finally the samples were recounted for 12 h after a 21–28 d decay for the remaining elements Ba, Ce, Co, Cr, Cs, Eu, Fe, Hf, Rb, Sb, Sc, Se, Th, U and Zn.

Photon activation can be used to determine 15 elements in soils using the internal standard method (Masumoto and Yagi, 1988). This requires the sample to be dissolved in acid and the residue mixed with the standard spike. The mixed sample and standard are then pressed into a pellet and irradiated with a linear accelerator for about 3 h with a beam current of 150 μA. The samples were measured at 3 h, 3 d and 1 month after irradiation. As a result 15 trace elements can be measured in calcareous loam soil and light sandy soil: Cr, Co, Ni, Zn, As, Rb, Sr, Y, Zr, Nb, Sb, Cs, Ba, Ce and Pb, in concentrations from 0.7–556 mg kg^{-1}, for Sb and Sr, respectively.

HUMUS, MULL AND TILL

Closely related to soil samples are the materials called humus, mull or till. They are basically the rotten leaves that form peat and contain proportions of soil in them. They are used for exploration purposes because of the ease of collection and the fact that some elements are concentrated in them. Rosenberg et al. (1983) have used epithermal neutron activation analysis for the routine analysis of till samples. This is an example of the beneficial use of epithermal neutron activation. A thermal neutron filter will reduce the activation of Sc, Fe, Cu, Na and actually improve the detection limits for a number of elements analyzed routinely such as Co, Ni, Rb, Zn, Sm, Yb, and Lu, as well as providing data for particular elements of interest such as gold. Samples weighing 0.1–1 g, depending on the density, are loaded into capsules made of cadmium, inside aluminum containers. They are irradiated for 25–30 h in a thermal neutron flux of 1.2×10^{16} n m^{-2} s^{-1} with a cadmium ratio for gold of 2. The samples are counted for 20 min after a decay of 4–6 d. Typically 27 elements can be determined with the following detection limits (in mg kg^{-1}): Na 250, Ni 40, Rb 15, Cs 0.6, Lu 0.05, Th 0.4, Sc 0.5, Zn 100, Mo 1.5, Ba 80, Hf 3, U 0.3, Cr 40, As 1, Ag 3, La 1.5, Ta 0.5, Fe 2500, Se 4, Sn 100, Sm 0.05, W 2, Co 2.5, Br 0.6, Sb 0.1, Eu 2 and Au 0.003.

Epithermal neutron activation has perhaps been most valuable in its application to geological samples. Although there is an enormous range of matrices examined in geological problems, most of them have major background effects due to aluminum in the case of short irradiations, sodium for intermediate, and scandium, iron or cobalt for long irradiations in a thermal neutron flux. The original work of Steinnes (1971) showed that these interferences could be significantly reduced using a thermal neutron filter and it has become a standard procedure to apply the technique to enhance the detection of some elements.

However, it is important to remember that only some elements will be improved and it is unusual for the method to be applied on a routine basis for multielemental analysis. It will be used in addition to thermal neutron activation to provide data for a few extra elements or as the chosen procedure for a specific analysis, such as in the case of gold determination. There will be no enhancement of the elements routinely measured by the short irradiation such as Ti, V, Mn, Mg, Cl, so the only advantage is that a sample 30 times larger gives the same radioactive dose rate on irradiation. However, the detection limits for elements with small cadmium ratios, such as Dy, Rh, Ag, Hf, Sc, U, Th, will be improved, as demonstrated by Parry (1982).

SEDIMENTS

Thirty-five elements were measured in a deep-sea sediment as part of a study on the disposal of highly radioactive waste in sub-seabed burial (Fong and Chatt, 1987). Samples of freeze-dried sediment weighing about 250 mg were sealed in polyethylene containers and irradiated in a thermal neutron flux of 5×10^{15} n m^{-2} s^{-1}. The full irradiation scheme is shown in Figure 15.2. The samples were irradiated for 4 min in a cadmium-lined pneumatic irradiation system, then counted for 10 min after a 7 min decay for Al, Ba, Br, Ca, Cl, I, Mg, Mn, Na, Si, Sr, Ti, U and V. Then the samples were irradiated for 7 h in the thermal neutron flux and counted for 30 min after a 3–4 d decay on both a lithium-drifted germanium detector and a low energy photon detector. Elements measured included As, Br, K, La, Sb, Sm, Ta, U, and Yb. After a 25 d decay they were counted again for 12 h on both detectors for Ba, Ce, Co, Cr, Cs, Eu, Fe, Gd, Hf, In, Lu, Nd, Rb, Sb, Sc, Se, Sr, Ta, Tb, Th, Tm, Yb, Zn and Zr.

Miller (1987) has analyzed samples of sandstones and shales as part of a study of a geothermal field in California. Drill-core samples were washed and dried, then broken up and ground in an agate mortar and pestle to 100 mesh. Samples weighing 0.5 g were taken for analysis. Three irradiations were made with the reactor at 1 kW, 15 kW and 250 kW for short, intermediate and long-lived radionuclides. The short irradiation

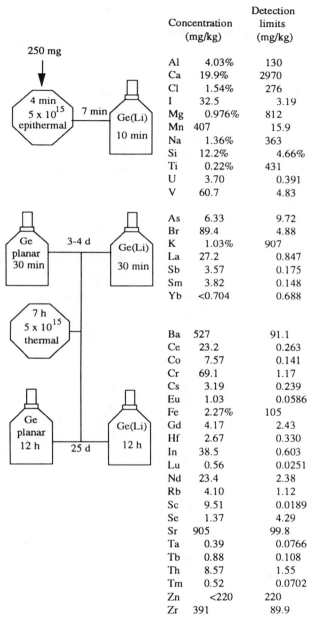

	Concentration (mg/kg)	Detection limits (mg/kg)
Al	4.03%	130
Ca	19.9%	2970
Cl	1.54%	276
I	32.5	3.19
Mg	0.976%	812
Mn	407	15.9
Na	1.36%	363
Si	12.2%	4.66%
Ti	0.22%	431
U	3.70	0.391
V	60.7	4.83
As	6.33	9.72
Br	89.4	4.88
K	1.03%	907
La	27.2	0.847
Sb	3.57	0.175
Sm	3.82	0.148
Yb	<0.704	0.688
Ba	527	91.1
Ce	23.2	0.263
Co	7.57	0.141
Cr	69.1	1.17
Cs	3.19	0.239
Eu	1.03	0.0586
Fe	2.27%	105
Gd	4.17	2.43
Hf	2.67	0.330
In	38.5	0.603
Lu	0.56	0.0251
Nd	23.4	2.38
Rb	4.10	1.12
Sc	9.51	0.0189
Se	1.37	4.29
Sr	905	99.8
Ta	0.39	0.0766
Tb	0.88	0.108
Th	8.57	1.55
Tm	0.52	0.0702
Zn	<220	220
Zr	391	89.9

Figure 15.2. Neutron activation analysis of a deep-sea sediment. (Adapted with permission from Fong and Chatt, 1987.)

201

was made in a thermal neutron flux of 6×10^{13} n m^{-2} s^{-1} for 30 s. The samples were counted for 4 min after a decay of 4 min, for Al, Ca, Ti and V. The second irradiation was for 15 min at 9×10^{14} n m^{-2} s^{-1}. After a decay of 90 min the samples were counted for 2 min for Mn and Dy and then 24 h later for Na, K and Sm. The third long irradiation was for 5 h at 1.5×10^{16} n m^{-2} s^{-1}. This was followed by three measurements after 5 d for 30 min for U and As; after 2 weeks delay for 1 h for Ba, Nd, La, Lu, Sr; and after 4 weeks for 2 h for Ag, Ce, Co, Cr, Cs, Eu, Fe, Hf, Rb, Sc, Se, Ta, Tb, Th, Zn and Zr.

Seabed analysis has also been carried out *in situ* using a californium neutron source and germanium detector (Wogman, 1977). The 1 mg californium source was used for a 2 min irradiation followed by a 40 s delay and a count for 2 min. Up to 33 elements were analyzed this way including the determination of Hf, In, Sc and Se down to a few mg kg^{-1}.

MARBLE

Most applications concerned with the analysis of marble are in archaeological studies. It is possible, with cluster analysis, to characterize the marble quarries on the basis of their trace element content and to identify the provenance of marble with which the monuments and artifacts were made. A study was made, for example, of the provenance of the white marble of the cathedral of Como (Oddone et al., 1985). Samples were irradiated for 1 h in a thermal neutron flux of 10^{16} n m^{-2} s^{-1} and counted after 3 h for U and Mn. The remaining elements, Sc, Cr, Fe, Co, Rb, Sb, Cs, Ba, La, Ce, Sm, Eu, Gd, Ho, Tm, Yb, Lu, Hf, Th and U were measured after a 60 h irradiation and a 1–3 week decay period. A similar study on white marble from Italy and Turkey was based on a single irradiation procedure followed by three measurements to determine 30 elements (Moens et al., 1988). The samples were irradiated for 7 h in a thermal neutron flux of 1.5×10^{16} n m^{-2} s^{-1}. They were counted for 15 h after 1 d, for 24 h after 1 week and for 24 h after 5 weeks. The elements detected were Na, K, Sc, Cr, Mn, Fe, Co, Zn, Ga, As, Br, Sr, Sb, Cs, Ba, La, Ce, Sm, Eu, Tb, Ho, Yb, Hf, Ta, Th and U. Atomic absorption spectrometry was used to provide supplementary data for aluminum, magnesium and vanadium.

SILICATE ROCKS

There are a wide variety of problems related to geochemistry. One of the advantages of activation analysis is the small sample size required for a measurement and so mineral separates which can weigh only a few milligrams may be used. Precious samples such as lunar material from the Apollo missions have been analyzed by neutron activation analysis. Generally the requirements in geochemistry fall into three types. The first is a multielement analysis where the total elemental composition of a sample is required; the second is the evaluation of the rare earth elements as a group to elucidate the processes during fractionation; thirdly there will be specific elements of interest such as the halides.

Multielement analysis can be applied to rocks and minerals in exactly the same way as for sediments and soils. There are a large number of typical schemes and an example is shown in Figure 15.3. It is the routine procedure used for the analysis of Chinese standard rocks for 35 elements (Sun and Jervis, 1985). Samples weighing 100–200 mg are irradiated in a Slowpoke® reactor with a thermal neutron flux of 2.5×10^{16} n m^{-2} s^{-1}. The short-lived radionuclides are measured following a 1 min irradiation and a 10–15 min decay. The samples are counted for 5 min for Al, Ba, Ca, Cl, Dy, I, Mg, Mn, Sr, Ti, U and V. A second irradiation for 16 h is followed by a 4–5 d decay before counting for 30 min for As, Br, K, La, Na, Sm and W. Finally the samples are recounted after 15–20 d for 100 min to measure the long-lived radionuclides of Ce, Co, Cr, Cs, Eu, Fe, Hf, Lu, Nd, Rb, Sb, Sc, Se, Sn, Ta, Tb, Th, Yb and Zn.

Neutron activation is used very widely for the determination of the rare earth elements (REE) for the study of geological processes. The similar chemical behaviour of the REE stems from the similarity of their ionic radius and charge (3+). Therefore if the concentrations of the REE are normalized by dividing by the corresponding chondritic values, a smooth pattern is obtained. However, because Eu is also able to exist in the 2+ state it may not behave in the same way as the rest of the REE. The chondrite normalized plot may therefore show an anomaly at europium, where the concentration may be higher or lower than expected. The result is a positive or negative peak which relates to the oxygen fugacity when the REE were incorporated into the crystal structure. Cerium may in addition show anomalous results since it can also exist in the 4+ state. To obtain information from the europium anomaly it is necessary to have precise data on both samarium and gadolinium values as well as europium. In the case of ^{153}Gd, where the interference is from ^{153}Sm, the samarium is allowed to decay before gadolinium is counted. Holmium and thulium are also subject to interferences which makes the data unreliable. A particular problem for the REE is the interference due to other radionuclides produced on fission of uranium in the sample during irradiation. In particular it can affect the determination of

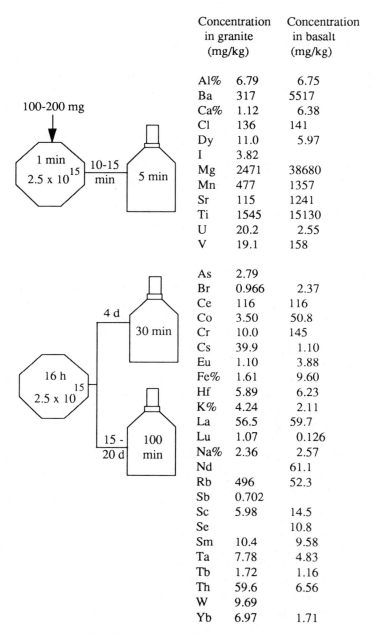

	Concentration in granite (mg/kg)	Concentration in basalt (mg/kg)
Al%	6.79	6.75
Ba	317	5517
Ca%	1.12	6.38
Cl	136	141
Dy	11.0	5.97
I	3.82	
Mg	2471	38680
Mn	477	1357
Sr	115	1241
Ti	1545	15130
U	20.2	2.55
V	19.1	158
As	2.79	
Br	0.966	2.37
Ce	116	116
Co	3.50	50.8
Cr	10.0	145
Cs	39.9	1.10
Eu	1.10	3.88
Fe%	1.61	9.60
Hf	5.89	6.23
K%	4.24	2.11
La	56.5	59.7
Lu	1.07	0.126
Na%	2.36	2.57
Nd		61.1
Rb	496	52.3
Sb	0.702	
Sc	5.98	14.5
Se		10.8
Sm	10.4	9.58
Ta	7.78	4.83
Tb	1.72	1.16
Th	59.6	6.56
W	9.69	
Yb	6.97	1.71

Figure 15.3. Neutron activation analysis of Chinese standard rocks. (Adapted with permission from Sun and Jervis, 1985.)

204

lanthanum, cerium and neodymium, by actually producing the radio-nuclides during the fission process. Corrections can be made by irradiating uranium of natural isotopic abundance and analyzing the resulting spectrum for interferences. There are a number of sources of interferences which are well documented by Potts (1987).

Many of the gamma rays of the REE are at quite low energies and in most cases it is of advantage to use a low energy photon detector. Planar germanium detectors are suited to the measurement of X-rays and low energy gamma rays since the resolution at low energies is superior to that of a large crystal. They are designed with a thin window of beryllium to avoid attenuation of the low energy rays. Labrecque and Rosales (1986) designed a procedure for the analysis of 500 mg samples using a single irradiation and a series of three measurements with a low energy photon detector. The sample was irradiated for 5 h at 5×10^{16} n m^{-2} s^{-1}. It was counted for 750 s after a decay of 1 week for Sm, Ho, U and La; for 1500 s after 3 weeks for Yb, Nd, Ce and Lu; and after 7 weeks it was counted for 3000 s for Tm, Gd, Eu, Tb and Th. If the planar detector is used the data obtained will be limited to low energy gamma rays and there will be no values for Co, Sc and Fe. However, additional elements W, Ta, Hf, U and Th are usually detected and Au can be measured down to about 50 μg kg^{-1}.

Dysprosium has only short-lived radionuclides and a separate short irradiation is required to determine that element. Praseodymium and erbium have half-lives close to that of sodium and consequently they will only be detected after the interfering sodium activity has decayed. Kennedy and Touhouche (1987) have used pseudocyclic neutron activation for erbium with ^{167m}Er which has a half-life of 2.28 s. The procedure consisted of 25 cycles of 1.7 s irradiation, 1.1 s decay and 4.0 s count, with a 7 min wait between each cycle. The detection limit was 0.5 mg kg^{-1}. As a routine package it is usually adequate to measure only the long-lived nuclides.

Inductively coupled plasma atomic emission spectrometry (ICP–AES) is the most usual alternative technique with comparative sensitivities and capabilities. However, ICP does require sample dissolution and ion exchange separation of the rare earth elements before analysis. Activation spectrometry has the additional advantage that other elements such as U, Th, Ta and Hf are detected at the same time and epithermal neutron activation is particularly useful in improving the detection limits for these elements.

Some elements are not detected with neutron activation analysis at the concentrations usually found in igneous rocks. They include fluorine and lead. The behavior of fluorine is of considerable interest since it tends to be concentrated in the late stages of magmatic differentiation. The sensitivity of other techniques such as fast neutron activation are not good enough and wet chemical methods are time-consuming. Randle (1985) used a Dynamitron accelerator with an integrated flux of 3×10^{10} n m^{-2} to analyze 5 g samples of rock for fluorine via the $^{19}F(n,\alpha)^{16}N$ reaction. The detection limit for fluorine was 1 mg kg^{-1}. Proton-induced gamma ray emission techniques have been used to measure fluorine in rock samples based on the $^{19}F(p,\alpha,\gamma)^{16}O$ reaction (Roelandts et al., 1986). Proton activation with a 3 MeV Van de Graaff accelerator using a pellet of rock powder mixed with graphite will give a detection limit of about 25 mg kg^{-1} for a sample weighing 180 mg. Proton activation has also been used to advantage for the determination of lead (Wauters et al., 1987). The detection limit in geological samples was 100–500 mg kg^{-1}, although 10 μg kg^{-1} could be detected if a radiochemical separation was used.

The fluid inclusions found in hydrothermal minerals can vary widely in chemical composition and in favorable conditions they can be analyzed instrumentally with neutron activation. Quartz provides a very clean matrix for determination of dissolved elements and detection limits can be very low. Luckscheiter and Parekh (1979) analyzed the quartz samples in the form of cylinders of 3 mm diameter and 10 mm high weighing about 2 g. They were irradiated for about 15 min in a neutron flux of 7×10^{16} n m^{-2} s^{-1} to determine Na and Cl. Later the samples were re-irradiated for 2–3 h to measure Na, K, Br, As and Mn. In this case it was possible to measure concentrations (in mg kg^{-1}) of Na 0.6, Cl 0.9, Br 0.1, Mn 0.002, As 0.03, and K 1.2.

URANIUM ORE

In mineral exploration there is less desire to determine a whole suite of elements and usually it is adequate to measure the chosen element or elements down to a certain predetermined limit. The main problem in the case of mineral exploration is that of representative sampling. Activation analysis has tended to be limited to small sample sizes, which is normally considered an advantage. Some pneumatic irradiation systems are designed for 1 g samples. This is simply insufficient material for a representative analysis in the case of many elements particularly the platinum group elements and gold.

Uranium is perhaps the best example of routine exploration by activation analysis with the use of the very specific method of delayed neutron counting to determine only uranium (Amiel, 1981). The technique, which has been in use for many years, is an excellent example of the use of activation analysis for a specific purpose. The operation can be readily automated and there are several examples of systems which operate completely unattended (Rosenberg et al., 1977). Samples weighing 10 g are packed in polyethylene containers and irradiated for 60 s in a flux of 10^{16} n m^{-2} s^{-1} then counted for 60 s after a decay of 20 s. The decay period is necessary to allow interfering neutrons from ^{17}N to decay. The detection limit for a standard rig containing six proportional counters is 0.6 μg of uranium, which corresponds to 0.06 mg kg^{-1}. It is possible to obtain better detection limits with a more sensitive counting system but for exploration purposes a detection limit of 0.1 mg kg^{-1} is adequate. The method can be used to analyze stream sediments and water samples if required.

Additional elements can be obtained with the same irradiation when delayed neutron counting is used. Shenberg et al. (1987) measured F, Al, Ca and V while the sample was waiting to be analyzed by delayed neutron counting. The samples of phosphatic material, and some oil shales, were irradiated in a thermal neutron flux of 5×10^{16} n m^{-2} s^{-1} for 20 s. They were then transferred to a gamma ray counting system where they were counted for 35 s live time during the following 60 s. Values were obtained for the concentrations of F, Al, Ca and V. The samples were then transferred to a delayed neutron counting rig to be counted for 60 s after the 60 s delay. The additional elements provided valuable information for the uranium survey, in particular the strong correlations observed for vanadium.

Uranium can also be determined via the ^{238}U(n,γ)^{239}U reaction. High concentrations of uranium can be measured after just a few minutes irradiation, as ^{239}U, which has a half-life of 22 min. Uranium has a low cadmium ratio and it is an example of an element which is enhanced with the use of epithermal activation. The gamma ray energy is at 74 keV and the use of a planar, low energy detector improves the detection limit. It is also possible to measure the decay product ^{239}Np which has a half-life of 2.6 d, with the usual routine multielement analysis package which includes the rare earth elements.

GOLD ORE

Gold is an element with one of the best sensitivities for neutron activation analysis. The abundance of the isotope is 100% and the ^{197}Au(n,γ)^{198}Au reaction has a high cross section. It is possible to measure concentrations down to 10 μg kg^{-1} using routine procedures. The radionuclide has resonances in the epithermal neutron region and therefore detection is enhanced by irradiation under a thermal neutron filter. Gold particles can be distributed heterogeneously in the sample. Although reducing the particle size helps, the homogeneity cannot always be improved by grinding due to the smearing of metallic gold. Consequently it is essential to analyze a sufficiently large sample and fire assay is used by the mining industry to measure gold routinely. Activation laboratories sometimes use fire assay preconcentration into a silver bead for representative analysis. The alternative is to analyze large samples, up to 50 g, instrumentally. The two problems associated with analyzing large samples, neutron self-shielding and gamma ray absorption effects, cannot be overcome without detailed knowledge of the samples. However, the requirements of mineral exploration may mean that in some cases the resulting errors are acceptable.

There is an alternative reaction of gold, $^{197}Au(n,n')^{197m}Au$, which has been implemented for analysis of gold. The ^{197m}Au has a half-life of 7 s and so the measurement is rapid. It is not very sensitive, despite the enhancement in fast neutrons and detection is limited to about 1 mg kg^{-1}. Cumulative activation, where several samples are irradiated in sequence, has been implemented to allow determination of 10 g samples (Parry, 1987). The activation has been optimized using an accelerator in Norilsk (USSR) which processes 250,000 samples per year (Burmistenko et al., 1977).

Vegetation has been used for geobotanical prospecting, particularly for gold. Cohen et al. (1987) collected the bark, leaves and twigs from spruce, alder, balsam, birch and mountain maple trees growing over areas of gold mineralization. Samples weighing 20 g were air dried in the field and then at 90° C for 12 h before milling. The milled samples were pressed into 8 g diskettes and grouped in bundles of 20 with replicates and an internal standard. The group of samples was irradiated for 1 h in a neutron flux of 7×10^{16} n m^{-2} s^{-1}, and counted for 100 s after a 4 d decay. The detection limit for gold was 0.05 μg kg^{-1} and 30 other elements were measured simultaneously. The work showed that gold is accumulated in the bark and leaves of balsam fir, spruce, alder, mountain maple and birch.

PLATINUM GROUP MINERALS

The platinum group elements (PGE) are not normally detected during the routine measurement of rock samples. The detection limits in rock samples are around several mg kg^{-1} and measurement is only possible in meteorites or ore samples with very high concentrations. The exception to that is iridium where INAA has been carried out on samples using the 468 keV gamma ray energy of ^{192}Ir.

The PGE, like gold, are usually dispersed heterogeneously in mineral phases and so large representative samples are analyzed. In general any analytical technique which is concerned with the determination of the PGE involves a preconcentration stage, usually fire assay and increasingly the collection into nickel sulfide bead (Hoffman et al., 1982). The bead is then crushed and dissolved and the insoluble PGE collected on filter paper prior to digestion, or in the case of activation analysis, irradiation as a solid.

Once the PGE are preconcentrated, neutron activation is a very sensitive technique for determination of the PGE. The samples are first irradiated for a short time, about 5 min, to activate the Rh and Pd. The use of epithermal neutron activation will enhance the detection. The detection limit is about 1 μg kg^{-1} for Rh and 10 μg kg^{-1} for Pd, based on a 50 g sample. The sample is then irradiated for several hours to activate the remaining elements and counted after about 24 h decay for Pt and Au and then measured again after about a week for Ir, Ru and Os.

Recent work to reduce the size of the nickel sulfide bead used to collect the PGE has resulted in the possibility of using activation spectrometry for the direct analysis of the bead (Asif and Parry, unpublished data). The detection limits are dependent on the copper content of the sample and in some cases concentrations of below 100 μg kg^{-1} can be achieved for all the PGE.

BOREHOLE LOGGING

Borehole logging is a neutron activation method for the *in situ* analysis of elemental concentrations in mineral exploration. The topic has been reviewed by Clayton (1977). The technique is limited by the problems of using a steady state neutron probe, which results in very complex spectra. Now that it is possible to use high resolution detectors in boreholes it is easier to evaluate the spectra and measurements have been made using a californium source and germanium detector. Normally the prompt gamma rays emitted by elements such as iron, copper, manganese, titanium, silicon, calcium, oxygen, potassium and aluminum are measured. Examples of borehole logging include the measurement of copper (Eisler et al., 1971), nickel (Senftle et al., 1971), and iron (Eisler et al., 1977).

REFERENCES

Adepetu, J. A., O. I. Asubiojo, F. Y. Iskander, and T.L. Bauer (1988), "Elemental composition of Nigerian Harmattan dust," *J. Radioanal. Nucl. Chem.*, **121**(1), 141–147.

Amiel, S. (1981), "Neutron counting in activation analysis," in S. Amiel (ed.), *Nondestructive Activation Analysis*, Elsevier, Amsterdam, pp. 43–52.

Burmistenko, U. N., V. A. Glukhikh, and I. N. Ivanov (1977), "Kompleks apparatury dlya fotoyadenogo opredeleniya zolota i soputstvuyushikh elementov v rudnykh probakh," in *Tezisy dokladov IV Vsesoyuznogo soveshchaniya po aktivatsionnomu analizu*, Tbilisi.

Clayton, C. G. (1977), "Some recent applications of nuclear techniques in the exploration and mining of metalliferous minerals," *Nuclear Techniques and Mineral Resources. Proc. Symp. Vienna*, IAEA–SM–216/102, International Atomic Energy Agency, Vienna, pp. 185–213.

Cohen, D. R., E. L. Hoffman, and I. Nichol (1987), "Biogeochemistry: a geochemical method of gold exploration in the Canadian Shield," *J. Geochem. Explor.*, **29**, 49–73.

Cornelis, R., J. Hoste, A. Speecke, C. Vandecasteele, J. Versieck and R. Gijbels (1976), "Activation Analysis – Part 2," in T. S. West (ed.), *Physical Chemistry Series Two, International Review of Science*, Butterworths, London, pp. 73–95.

Das, H. A., A. Faanhof, and H. A. van der Sloot (1989), *Radioanalysis in Geochemistry*, Elsevier Science Publishers, Amsterdam.

Eisler, P. L., P. Huppert, and A. W. Wylie (1971), "Logging of copper in simulated boreholes by gamma spectroscopy: 1. Activation of copper with fast neutrons," *Geoexploration*, **9**, 181–194.

Eisler, P. L., P. J. Mathew, A. W. Wylie and S. Youl (1977), "Use of neutron capture gamma radiation for determining grade of iron ores in blast and exploration holes," *Nuclear Techniques and Mineral Resources. Proc. Symp. Vienna*, IAEA–SM–216/3, International Atomic Energy Agency, Vienna.

Fong, B. B. and A. Chatt (1987), "Characterization of deep sea sediments by INAA for radioactive waste management purposes," *J. Radioanal. Nucl. Chem.*, **110**(1), 135–145.

Hoffman, E. L., A. J. Naldrett, J. C. Van Loon, R. G. V. Hancock, and A. Manson, "The determination of all the platinum group elements and gold in rocks and ore by neutron activation analysis after preconcentration by a nickel sulphide fire-assay technique on large samples," *Anal. Chim. Acta*, **102**, 157–166.

Kennedy, G. and K. Touhouche (1987), "Determination of erbium in silicate rocks at the 1 μg/g level", *J. Radioanal. Nucl. Chem.*, **114**(2), 319–327.

Labrecque, J.J., and P.A. Rosales (1986), "Application of an Apple IIe microprocessor for data acquisition and analysis in instrumental neutron activation analysis with a low energy photon detector," *J. Radioanal. Nucl. Chem.*, **102**(2), 377–384.

Luckscheiter, B. and P. P. Parekh (1979), "A new method for the determination of dissolved elements in fluid inclusions," *N. Jb. Miner. Mh.*, **3**, 135–144.

Masumoto, K. and M. Yagi (1988), "Revaluation of the internal standard method coupled with the standard addition method applied to soil samples by means of photon activation," *J. Radioanal. Nucl. Chem.*, **121**(1), 131–139.

Miller, G. E. (1987), "Neutron activation analysis of core and drill cutting samples from geothermal well drilling," *J. Radioanal. Nucl. Chem.*, **114**(2), 351–358.

Moens, L., P. Roos, J. De Rudder, J. Hoste, P. De Paepe, J. Van Hende, R. Marechal, and M. Waelkens (1988), "White marble from Italy and Turkey: an archaeometric study based on minor- and trace-element analysis and petrography," *J. Radioanal. Nucl. Chem.*, **123**(1), 333–348.

Oddone, M., S. Meloni, and E. Mello (1985), "Provenance studies of the white marble of the Cathedral of Como by neutron activation analysis and data reduction," *J. Radioanal. Nucl. Chem.*, **90**(2), 373–381.

Parry, S. J. (1982), "Epithermal neutron activation analysis of short-lived nuclides in geological material," *J. Radioanal. Chem.*, **72**(1–2), 195–207.

Parry, S. J. (1987), "Cumulative neutron activation of short-lived radionuclides for the analysis of large samples in mineral exploration,"*J. Radioanal. Nucl. Chem.*, **112**(2), 383–386.

Potts, P. J. (1987), *A Handbook of Silicate Rock Analysis*, Blackie, Glasgow, pp. 424–429.

Randle, K. (1985), "Determination of fluorine in geological samples using accelerator derived neutrons," *J. Radioanal. Nucl. Chem.*, **90**(2), 355–361.

Robotham, H., G.C. Lalor, A. Mattis, R. Rattray, and C. Thompson (1987), "Trace elements in Jamaican soils: 1. The parishes of Clarendon, St. Catherine, Portland, and St. Elizabeth," *J. Radioanal. Nucl. Chem.*, **116**(1), 27–34.

Roelandts, I., G. Robaye, G. Weber and J. M. Delbrouck-Habaru (1986), "The application of proton-induced gamma-ray emission (PIGE) analysis to the rapid determination of fluorine in geological materials," *Chem. Geol.*, **54**, 35–42.

Rosenberg, R. J., V. Pitkanen, and A. Sorsa (1977), "An automatic uranium analyser based on delayed neutron counting," *J. Radioanal. Chem.*, **37**, 169–179.

Rosenberg, R., R. Zilliacus and M. Kaistila (1983), *Neutron Activation Analysis of Geochemical Samples*, Technical Research Centre of Finland Research Notes 225, Espoo, Finland.

Salmon, L. and P. A. Cawse (1983), "Instrumental neutron activation analysis," in K. A. Smith (ed.), *Soil Analysis*, Marcel Dekker, New York, pp. 299–353.

Senftle, F. E., P. F. Wiggins, D. Duffey, and P. Philbin (1971), "Nickel exploration by capture gamma rays," *Econ. Geol.*, 66, 583–590.

Shenberg, C., Y. Nir-El, Z. Alfassi, and Y. Shiloni (1987), "Rapid and simultaneous determination of U, F, Al, Ca and V in phosphate rock by a combination of delayed neutron and γ-ray spectrometry techniques," *J. Radioanal. Nucl. Chem.*, **114**(2), 367–377.

Smith, K. A. (ed.) (1983), *Soil Analysis*, Marcel Dekker, New York.

Steinnes, E. (1971), "Epithermal neutron activation analysis of geological material," in A. O. Brunfelt and E. Steinnes (eds), *Activation Analysis in Geochemistry and Cosmochemistry*, Universitetsforlaget, Oslo, pp. 113–128.

Sun, J. X., and R. E. Jervis (1985), "Neutron activation analysis of 35 elements in Chinese standard rocks (GSR) and soils (GSS) using the Slowpoke reactor," *J. Radioanal. Nucl. Chem.*, **89**(2), 553–560.

Wauters, G., C. Vandecasteele, and J. Hoste (1987), "The determination of lead in environmental and geological materials by proton activation analysis," *J. Radioanal. Nucl. Chem.*, **110**(2), 477–490.

Wogman, N. A. (1977), "In-situ X-ray fluorescence and californium-252 neutron activation analysis for marine and terrestrial mineral exploration," *Nuclear Techniques and Mineral Resources. Proc. Symp. Vienna*, IAEA–SM–216/56, International Atomic Energy Agency, Vienna.

INDUSTRIAL APPLICATIONS

Industrial applications of NAA cover an extraordinary range of problems. One of the main interests is the determination of the purity of material for quality control or to study the role of the impurities on the product or production (Dams et al., 1976). In these cases it is necessary to ensure that impurities are below a specified detection limit, and with material such as semiconductor silicon these detection limits may be set very low. It is difficult to categorize industrial samples since they can cover a very wide range of material, and some of them will have been discussed in earlier chapters on geological and environmental applications. In this chapter an attempt has been made to collect similar materials under one heading, according to the type of matrix rather than for a particular type of industry, which could include very diverse materials.

CARBON AND BORON

Carbon presents an interference free matrix for neutron activation analysis. Since there is little activation of the matrix the sample size can be in grams without neutron self-shielding and, because the atomic number is low, there is no problem from gamma ray attenuation. Whatever the analytical problem the conditions for irradiation and counting can be selected to optimize determination of the element of interest since the matrix has no effect on the measurement. Consequently any of its forms, carbon fiber, graphite, activated charcoal and diamond, may be analyzed very easily and concentrations of trace elements below mg kg^{-1} are determined in most cases.

Activation analysis of carbon is used to solve a variety of problems. It may be the purity of the material itself that is of interest, as in the case of carbon fiber used for a reactor irradiation tube (Bode and de Bruin, 1988). Carbon rod of the type used in atomic absorption spectrometry can be analyzed without pretreatment. It may, in the form of activated charcoal, have been used to collect an element of interest, such as gold. It can be used as a catalyst support, in which case it may be loaded with per cent levels of platinum or rhodium. Carbon in the form of blocks or fiber are not very easily handled but can be cleaned up after irradiation if necessary. Even diamonds, which are very difficult to analyze by other techniques, have been evaluated for trace impurities by neutron activation (Fesq et al., 1973). It is possible to identify and measure about 40 elements as impurities, some in concentrations below one μg kg^{-1}.

Boron is also a very difficult element to dissolve and analyze by other techniques. It does not emit gamma rays when irradiated with a neutron flux and therefore trace elements are easily detected. The major problem in the analysis of boron is that it absorbs thermal neutrons, which attenuates the neutron flux in the sample, making the standardization of the measurement difficult. Davies et al. (1986) overcame the problem by irradiating the sample in a boron filter to exclude the thermal neutrons. The irradiation scheme is given in Figure 16.1 with typical values and the detection limits. Samples weighing 0.15 g were irradiated in silica tubes inside a boron carbide liner with an effective cutoff to 180 eV. Up to 60 trace elements were determined in boron and its compounds, with limits of detection for most elements at low mg kg^{-1} or below. The detection limits were extremely good in this particular case since the material was very pure. If substantial amounts of impurities are present, the background from them will affect the detection limits of other trace elements.

PETROLEUM, OIL AND POLYMERS

Since the elements carbon, oxygen and hydrogen do not activate significantly on irradiation, materials such as polyethylene, polyurethane and polymethylmethacrylate are very suitable for trace element analysis with neutron activation. Activation spectrometry has a major advantage over other techniques which require a liquid sample, since digestion of these materials could result in losses of trace elements. In addition the trace element concentrations are very low in some plastics, below the mg kg^{-1} level, and it would be difficult to dissolve and analyze the sample without possible contamination. The main problem that confronts the activation analyst is keeping the material free from contamination before it is irradiated. Since this is almost impossible it is preferable to irradiate the sample and then remove surface contamination after irradiation by washing or even grinding off the surface, provided that there is time available before counting.

Twenty elements were measured in polyethylene used for making

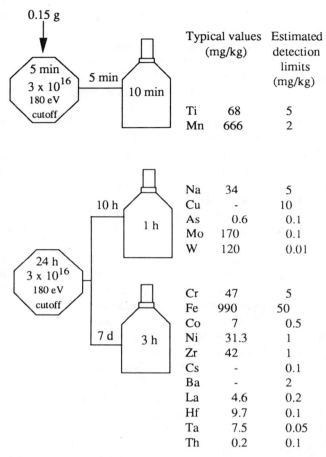

| Typical values (mg/kg) | Estimated detection limits (mg/kg) |

	Typical values (mg/kg)	Estimated detection limits (mg/kg)
Ti	68	5
Mn	666	2
Na	34	5
Cu	-	10
As	0.6	0.1
Mo	170	0.1
W	120	0.01
Cr	47	5
Fe	990	50
Co	7	0.5
Ni	31.3	1
Zr	42	1
Cs	-	0.1
Ba	-	2
La	4.6	0.2
Hf	9.7	0.1
Ta	7.5	0.05
Th	0.2	0.1

Figure 16.1. Epithermal neutron activation analysis of boron. (Adapted with permission from Davies et al., 1986.)

irradiation containers (Kucera and Soukal, 1983). Samples weighing between 300 and 500 mg were irradiated in a thermal neutron flux of $2\text{--}3 \times 10^{17}$ n m^{-2} s^{-1}. A short irradiation of 3 min, followed by a 4 min decay and a count for 5 min was used to measure Al and V. A 5 min irradiation was followed by a 1 h decay before counting for 15 min for Cl and Mn. A 10 h irradiation was followed by a 2 d decay before counting for 30 min for As, Au, Br, Cu, Hg, K, La, Mo, Na, Sb and Sm. The samples were recounted after 24 d, for 2 h, to measure Co, Cr,

Fe, Hg, Sb, Sc and Zn. Detection limits are not quoted but typical values for polyethylene are in μg kg^{-1}.

The analysis of a material like polyethylene will give detection limits which are as close to perfect as possible. Other polymers such as polyurethane contain air which may give higher radioactivity due to argon. Some problems can occur with polymethylmethacrylate when the irradiation container can rupture due to evolution of gas resulting from radiation damage. Aardsma et al. (1987) had to restrict their irradiation of the polymer to 2 h in a neutron flux of 10^{17} n m^{-2} s^{-1} when they were determining thorium at the μg kg^{-1} level. To overcome the limitation on sensitivity because of the short irradiation time, they had to count the sample for 40 h to measure a concentration of 1 μg kg^{-1} of thorium.

Fast neutron activation can be used to determine antimony and chlorine in synthetic rubbers (Bild, 1982). Antimony and chlorine are incorporated as fire retardants into synthetic rubbers that are used to insulate electrical cables. New formulations are analyzed for these elements but any analytical method involving dissolution may result in the elements being lost due to their volatility. Activation analysis avoids these problems. The materials studied included a synthetic rubber made of chlorosulfonated polyethylene, and a copolymer of ethylene and propylene. The irradiations were performed in a 14 MeV neutron generator with titanium tritide targets. The neutron flux was 10^{11} n s^{-1}. The samples, 6 mm wide by 40 mm long, weighing 0.6–0.9 g were irradiated for 200 s and counted for 300 s after a 240 s decay period. Typical values obtained for a synthetic rubber were 2.4% Sb and 3.9% Cl.

Oil can be analyzed in a routine manner and Filby et al. (1985) have determined 20 elements by instrumental neutron activation. Samples weighing between 0.5 and 1.0 g are irradiated in a thermal neutron flux of 6×10^{16} n m^{-2} s^{-1}. The first irradiation is a short one, for 3–5 min. After 1 min decay the samples are counted for 150–180 s for Al, Mg, Ti and V. A second measurement after 30 min, for 1000 s, is used to determine Cl and Mn. The second irradiation is for 8 h, and the samples are allowed to decay for 36 h before counting for 1 h each for As, Br, Ga, K, La, Na and Sm. Finally, after 21 d the samples are recounted for 22 h each for Ba, Ce, Co, Cr, Eu, Fe, Hf, Hg, Ni, Rb, Sb, Sc, Se, Sr, Ta, Tb, Th and Zn.

Vanadium is determined routinely in petroleum using a californium-252 neutron source (Lubkowitz et al., 1980). The whole system is automated with pneumatic transfer for irradiation and counting. The neutron flux is low, 10^{12} n m^{-2} s^{-1}, but it provides a detection limit of 10 mg kg^{-1} for a sample weighing between 5 and 10 g. Six samples may be processed per hour and some 10,000 samples may be analyzed per year.

PHARMACEUTICALS AND COSMETICS

Pharmaceuticals such as aspirin have been analyzed with neutron activation, using the scheme in Figure 16.2 (Iskander et al., 1986). The samples, weighing 300–500 mg are irradiated in a thermal neutron flux of 2×10^{16} n m^{-2} s^{-1} for 3 min and counted for 3 min after a 1–2 min decay for Al, Mg, Ti and V; and after a 30–45 min decay for 15 min, for Mn and Cl. A second irradiation of 8 h is followed by a 24–36 h decay before a 1 h count for As, Br, K and Na. Finally, after a decay of 21–28 d, each of the samples is counted for 12 h for Ba, Ce, Co, Cr, Cs, Eu, Fe, Hf, Rb, Sb, Sc, Se, Sr, Th and Zn.

The active ingredients in drugs have been analyzed for purity by Kanias and Choulis (1985). In particular they were interested in the determination of I, Mn, Ag and Na in a quality control exercise. Iodine was determined by irradiating a 700 mm^3 sample for 5 min in a flux of 5.8×10^{14} n m^{-2} s^{-1} and counting for 80 s after a 30 min decay. Iron was measured in a sample weighing 200 mg, or 700 mm^3 by volume, which was irradiated in a flux of 2.9×10^{17} n m^{-2} s^{-1} for 5 min. The samples were counted for 1 h after a 5 d decay. The same neutron flux was used to determine manganese but samples weighing only 100 mg were irradiated for 6 min. They were counted for 3 min after a minimum decay of 2 h. A very short irradiation was used for silver, only 20 s at 1.2×10^{16} n m^{-2} s^{-1}, followed by a 3 min decay and a 40 s count. Finally a neutron flux of 5.8×10^{14} n m^{-2} s^{-1} was used for sodium, with a 5 min irradiation, a 1 h decay and a count of only 100 s. Typical values obtained for the different drugs were 0.5% Ag in eyedrops, 252 μg of Mn in capsules, 10% Na in powder for injection and 14% I in liquid injection.

Quality control measurements were also applied to the purity of large-volume parenteral solutions (Kanias, 1987a). These are active pharmaceutical products which contain amino acids, carbohydrates, calcium, chlorine, magnesium, potassium and sodium. They were counted in solution using a routine procedure to measure chlorine and sodium. Samples with a volume of 700 mm^3 were irradiated in a flux of 6×10^{15} n m^{-2} s^{-1} for 2 min and then 500 mm^3 was transferred to a clean container for counting. After no more than 4 min decay the samples were counted for 8 min. The total analysis time was only 20 min, including sample preparation time of 4 min, irradiation time of 2 min, a decay of

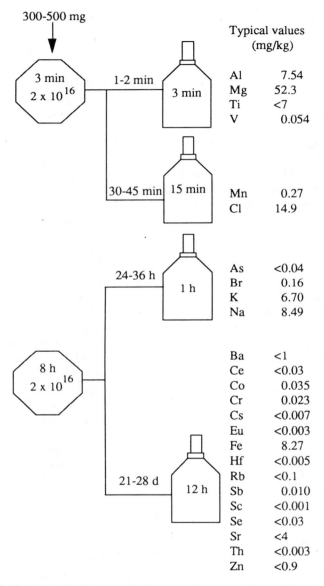

Figure 16.2. Neutron activation analysis of aspirin (Adapted with permission from Iskander et al., 1986.)

3 min, counting time of 8 min, and 3 min to calculate the result. The detection limits were 2.2 μg for Cl and 1.8 μg for Na.

Cosmetic products are controlled for purity in the same way as pharmaceuticals and Kanias (1987b) has used neutron activation to analyze eye shadow, face powder and rouge for zinc and iron. Samples weighing 200–300 mg were encapsulated and irradiated for 30 min in a thermal neutron flux of 2.7×10^{17} n m^{-2} s^{-1}. They were allowed to decay for 20 d before counting for 30 min. The routine procedure gave a determination limit of 45 μg of iron and 0.35 μg of zinc. Typical levels in eye shadow, depending on the color, ranged from 11% Fe and 769 mg kg^{-1} Zn in brown eye shadow to 0.448% Fe and 52 mg kg^{-1} Zn in pistachio color.

COAL AND METALLIFEROUS ORES

The role of neutron activation in the analysis of coal is described fully by Valkovic (1983) in a book on trace elements in coal. It provides detailed information on the technique including routine procedures, detection limits and comparisons with other techniques. The main components of coal are C, O, H, S, N, Si, Al and Fe. The aluminum and iron will be around 1% or less and so do not create particular difficulties for multielement analysis. Fast neutron activation can be used to determine the major elements and epithermal neutron activation has been applied to the determination of 22 elements in coal (Bellido and Arezzo, 1986). However it is possible to use thermal neutron activation for multielement analysis. Sun Jingxin and Jervis (1987) determined 35 trace and minor elements in coal and ash samples. Coal samples were irradiated for 5 min, and ash samples for 1 min, in a thermal neutron flux of 2.5×10^{15} n m^{-2} s^{-1}. After a 15 min decay period the samples were counted for Al, Ba, Ca, Cl, Dy, I, Mg, Mn, Sr, Ti, U, V and S. They were recounted after 3–5 d for As, Br, K, La, Na, Sm, W and Yb. A second irradiation of 16 h, with a 20 d decay, was used to measure Ce, Co, Cr, Cs, Eu, Fe, Hf, Lu, Nd, Rb, Sb, Sc, Se, Ta, Tb, Th and Zn.

Concentrations of 51 elements were determined in a reference coal using a combination of thermal neutron activation with prompt and delayed gamma ray spectrometry (Germani et al., 1980). Samples weighing about 100 mg were irradiated in a thermal neutron flux of $2–5 \times 10^{17}$n m^{-2} s^{-1}. A 5 min irradiation was made for Mg, Al, Cl, Ca, Ti, V, Mn, Br, Sr, In, I and Dy, with a count at between 6 and 12 min after a 15–20 min decay period. A 4 h irradiation was used to measure Sc, Cr, Fe, Co,

Zn, Se, Rb, Zr, Sb, Cs, La, Ce, Nd, Eu, Tb, Yb, Lu, Hf, Ta, Th and
U, with a 4 h count after 4–5 d, and a 8–24 h count after 20–30 d.
Intermediate half-life radionuclides of Na, K, Ga, As, Ba, Sm and W
were detected in both measurements. Prompt gamma ray analysis was
used to provide data for H, B, C, N, Na, Al, Si, S, Cl, K, Ca, Ti, Mn,
Fe, Cd, Nd, Sm and Gd.

A comparison of photon activation and thermal neutron activation
analysis of coal has shown that the methods are complementary (Pringle
et al., 1985). Photon activation is superior for Mg, Ni, Pb, Y and Zr,
which can be determined with a 0.5 g sample irradiated for 5–6 h and
counted for 3 h after a 1 d decay, plus a 10 h count following a 5 d
decay. The scheme used for thermal neutron activation consisted of a 5
min irradiation in a flux of 2.5×10^{15} n m^{-2} s^{-1} followed by a 5 min
count after 20 min. Then the samples were reirradiated for 16 h and
counted for 30 min after 3 d, and for 4 h after another 18 d. Elements
determined included As, Ba, Br, Ca, Ce, Co, Cr, Fe, I, Mn, Na, Rb,
Sb, Sr, Ti and U.

Fast neutron activation with a neutron generator with a flux of 5×10^{13}
n m^{-2} s^{-1} was used to determine percentage levels of sulfur in coal with
the $^{34}S(n,p)^{34}P$ reaction (Klie and Sharma, 1982). The half-life of the
product is 12.4 s and cyclic activation with 10 cycles of 1 min duration
were accumulated to gain a sensitivity of 0.25% S with a 5 g sample.
Sulfur can be measured on-line in lead sinter feed in order to optimize
the lead sinter machine performance (Cunningham et al., 1984). A
^{238}Pu–Be neutron source is used to irradiate the sample and the prompt
gamma rays emitted from neutron inelastic scattering are measured.
Determinations of S, Pb, Zn and Fe can be made to within 0.3, 0.9, 0.2
and 0.3 weight %, respectively.

Nuclear techniques applied to the coal industry are reviewed by Clayton
and Coleman (1986). A number of nuclear techniques developed by the
CSIRO Division of Mineral Physics, Australia for the bulk analysis of
coal and metalliferous ores have been reviewed by Borsaru et al. (1983).
Iron ore fines have been analyzed for iron and aluminum while moving
on a rubber conveyor belt (Holmes et al., 1980). A moderated ^{252}Cf
neutron source producing thermalized neutrons placed under the conveyor
belt is used to activate the iron which emits prompt gamma rays which
are measured with a sodium iodide detector above the irradiation site.
The ^{28}Al is measured with a second detector further down the belt.
Typical concentrations measured are 52.2–66.3% Fe and 1.3–6.7% Al
and the accuracy is quoted as 0.7% Fe and 0.1% Al. Kilogram quantities
of coal can be analyzed for aluminum and silicon with fast neutrons
(Borsaru and Mathew, 1982). A ^{241}Am-Be neutron source is used to

induce the $^{27}Al(n,p)^{27}Mg$ and $^{28}Si(n,p)^{28}Al$ reactions. The contents of the coals are in the ranges 1.3–10.3% alumina and 6.4–22% silica. A similar procedure is used to determine silica and alumina in bulk bauxite samples. Up to 10 analyses can be made per hour, using weights of about 3.5 kg. The samples analyzed contained 48–62% alumina and 2–11% silica (Borsaru and Eisler, 1981).

A high energy linear electron accelerator is used for photonuclear activation analysis of gold ore (Burmistenko et al., 1977). A system called Aura, based on an 8 MeV electron accelerator, is used for the inelastic scattering of photons in gold with the formation of the short-lived ^{197m}Au which has a half-life of 7.2 s. The analysis takes 40 s and the sampling capacity of the system at a mining site is 250,000 samples per year. The system can replace traditional fire assay laboratories with a detection limit of 0.4 mg kg^{-1} and only 10% error on a value of 2 mg kg^{-1}.

METALS AND ALLOYS

The ability to clean a sample after irradiation to remove any surface contamination is a major advantage in the determination of trace elements in pure materials. There is often a requirement to check the impurities in materials that are used for a wide range of applications, not least in the nuclear industry where aluminum, steel and Zircaloy of very high purity are required. Unlike the light elements and polymers most metals are activated to some extent by thermal neutrons and the problem is to measure a trace element in a material which contributes to an interfering background. Consequently the metals that are analyzed with neutron activation will be those with short-lived radionuclides, such as aluminum or rhodium, and those with long-lived radionuclides which do not activate well, such as zirconium. Elements like tungsten which do activate well cannot be analyzed for trace elements with neutron activation. There is an additional problem of neutron self-shielding and gamma ray attenuation due to the thickness of the sample. Measurements on a range of sample sizes will help illustrate whether the sample is suffering from these effects or not.

An example of the analysis of a metal which has a short-lived radionuclide is the determination of 11 trace elements in reactor grade aluminum used as fuel cladding (Pitrus, 1988). Samples weighing 100 mg were irradiated for 72 h in a thermal neutron flux of 7.6×10^{16} n m^{-2} s^{-1}. The surface of the sample was etched with nitric acid after irradiation to remove any surface contamination. It was then placed in a clean container and allowed to decay for between 5 and 20 days before counting for 2000 s to measure

Sc, Cr, Fe, Co, Zn, As, Sb, W, Au, Th and U. Detection limits were below mg kg^{-1} levels in all cases.

Zirconium is also used as a canning material in water-cooled reactors and it is a characteristic of the material that its corrosion resistance is impaired by a number of non-metallic impurities. Figure 16.3 shows the scheme used to determine elements with both short-lived and long-lived radionuclides in Zircaloys (Al-Jobori, 1988). Samples weighing between 2 and 200 mg were irradiated on four different occasions. The first was a 5 min irradiation in a low flux of 4.8×10^{15} n m^{-2} s^{-1}, and the sample was counted for 150 s after a 1 min decay for aluminum. A second irradiation in the same flux for 10 min was followed by a decay of 30 min before the sample was counted for 1000 s for manganese. A 6 h irradiation at a higher flux, 2.5×10^{17} n m^{-2} s^{-1}, was used to measure As and W after a 2 d decay and a count for 2000 s. Finally a 72 h irradiation was made in a flux of 3.0×10^{17} n m^{-2} s^{-1}. After 5 d the sample was counted for 2000 s for Sb and then after 11 d it was recounted for Co, Cr, Fe, Hf, Ni and Sn. Although it is really necessary to use radiochemical techniques to analyze tungsten, it is possible to determine Ba, Co, Cr, Fe, Ni, Sb, Sr, Ta and Th instrumentally where the concentrations are in the range of 10 to 100 mg kg^{-1} (Pitrus, 1988).

In the case of alloys, and of metals if they are not very pure, it may not be the material itself but the presence of impurities which activate well that causes a problem. For example, the manganese content of steel is sufficiently high to cause a major background even on a very short irradiation and prevents the determination of niobium in steel with the short-lived radionuclide of niobium. In the case of a long irradiation, chromium is the major activity in steel. Fast neutron activation has been used to determine nitrogen in steel via the ^{14}N(n,p)^{14}C reaction. A 20 min irradiation in a flux of 2.5×10^{18} n m^{-2} s^{-1} is used and the ^{14}C is separated afterwards. Photon activation analysis was also used and although it was less sensitive it also provided information about the carbon content at the same time. In this particular case it was most important to remove surface contamination and post irradiation cleaning could be used in either case. Fast neutrons have also been used to measure fluorine in steelmaking slag (Chiba, 1981). The reaction ^{19}F(n, p)^{19}O was used to avoid interferences from oxygen present in the steel. Samples were irradiated for 30 s in a neutron flux of 10^{10} n s^{-1} and counted for 15 s after a 60 s decay.

The composition of solder was used as forensic evidence in a case concerning a time device used in a bombing incident (Kishi, 1987). Fragments of the lead and tin solder were analyzed for As, Cu, Sb, Ag, Fe and Ni. The analysis of shotgun bullets is a well established technique

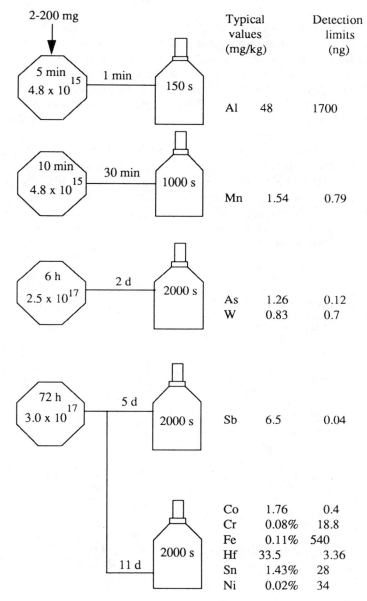

Figure 16.3. Neutron activation analysis of Zircaloy clad. (Adapted with permission from Al-Jobori, 1988.)

223

of identification in forensic work. The content of bullet lead varies according to the manufacturer and typically it contains up to per cent levels of Sb, As and Sn and mg kg^{-1} of Ag and Cu. Guinn et al. (1987) have developed a rapid method for the determination of all five elements. The shotgun pellet is flattened and a small disc is taken as a sample weighing between 20 and 30 mg. The samples are irradiated for 3 min in a thermal neutron flux of 2.2×10^{18} n m^{-2} s^{-1} and then counted for 3 min after a 5 min decay period.

There are several applications to archaeological studies involving the analysis of alloys, such as bronze. The example shown in Figure 16.4 (Holtta and Rosenberg, 1987) is the determination of the elemental composition of bronze and copper objects. Very small samples were taken from the objects by drilling with a 2 mm diameter steel drill. For large objects 60 mg samples were obtained but only 20 mg samples were taken from small objects. Instrumental analysis could only be carried out on

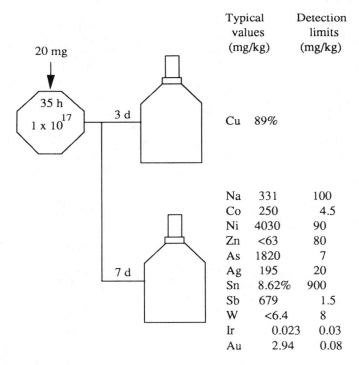

	Typical values (mg/kg)	Detection limits (mg/kg)
Cu	89%	
Na	331	100
Co	250	4.5
Ni	4030	90
Zn	<63	80
As	1820	7
Ag	195	20
Sn	8.62%	900
Sb	679	1.5
W	<6.4	8
Ir	0.023	0.03
Au	2.94	0.08

Figure 16.4. Neutron activation analysis of bronze axes. (Adapted with permission from Holtta and Rosenberg, 1987.)

those elements which have radionuclides with half-lives longer than that of ^{64}Cu (12.8 h) and so Mg, Al, V, Ti and Mn were measured after a radiochemical separation. Samples were activated for 35 h in a rotating irradiation device with a thermal neutron flux of 10^{17} n m^{-2} s^{-1}. The samples were dissolved in acid before counting to give reproducible geometry with a synthetic standard. The samples were then allowed to decay for 3 d, counted for copper, then re-analyzed after one week to determine Na, Co, Ni, Zn, As, Ag, Sn, Sb, W, Ir and Au.

The non-destructive nature of activation analysis is used to advantage in archaeological applications and in particular for coins, which may be analyzed whole. The determination of silver in silver coins can be done very rapidly without building up a lot of long-lived activity. Rosenberg (1985) irradiated ancient silver coins for only 4 s in a neutron flux of 4×10^{16} n m^{-2} s^{-1} and counted them after a few days for ^{110m}Ag. Neutron absorption corrections were determined by irradiating silver discs. Whole silver–copper coins 2.3 mm thick and 20 mm in diameter were irradiated in thermal neutrons for the determination of nine elements (Rousset and Federoff, 1985). The samples were irradiated for 5 s in a thermal neutron flux of 10^{17} n m^{-2} s^{-1}. They were counted for 15 min after a 30 min decay for Sn and In. After 3 h they were counted for 30 min for Mn and Ga and finally after 4 d they were recounted for 30 min for Cu, Au, As, Sb and Ag. Typical samples contained 57% Ag, 45% Cu, 90 mg kg^{-1} As, 577 mg kg^{-1} Sb, 6.8 μg kg^{-1} In, 1.1 mg kg^{-1} Mn, <2 mg/kg Ga and <0.3% Sn.

Fast neutrons have been implemented for the analysis of Gaulish silver alloyed coins (Elayi et al., 1985). A Van de Graaff accelerator with a 14 MeV neutron flux of 10^{10} n s^{-1} was used. Whole coins were irradiated for 5 min and counted after 3 min for Ag and Cu. Chalouhi et al. (1982) used a 14 MeV neutron generator with a flux of 10^{10} n s^{-1} to measure silver and copper via the ^{63}Cu(n,2n)^{62}Cu and ^{107}Ag(n,2n)^{106}Ag reactions. The half-lives of the products are 9.8 and 24 min respectively. Coins weighing up to 14 g were analyzed and the precision in the determination of copper and silver was better than 2% in recent samples, rather worse in ancient coins of irregular shape.

SEMICONDUCTOR MATERIALS AND GLASSES

Developments in the semiconductor industry have resulted in an increasing interest in the analysis of pure material by activation analysis. The characterization of inorganic materials used in the electronics industry is reviewed in a recent conference on the subject (Casper, 1986) with specific review of the role of activation analysis (Lindstrom, 1986) and neutron depth profiling (Downing et al., 1986). The main problems in semiconductors are caused by impurities of light elements, particularly carbon, nitrogen and oxygen. Such elements are not determined by neutron activation but can be measured using charged particles (Hoste and Vandecasteele, 1987).

A typical example of the analytical problems encountered is the microanalysis of oxygen and nitrogen in TiN thin films doped with oxygen, which are potentially useful in solar cell technology as antireflective and conductive layers. Berti and Drigo (1982) have used a 2 MeV Van de Graaff accelerator for deuteron-induced reactions on ^{16}O and ^{14}N which produce protons and alpha particles. Nitrogen and oxygen were measured simultaneously in silicon with a sensitivity of a few monolayers. Proton activation has been used to determine arsenic which is added as a dopant to high purity silicon in the production of shallow junctions (Birattari et al., 1987). Slices of monocrystalline silicon 500 μm thick were analyzed for arsenic using the $^{75}As(p,3n)^{73}Se$ reaction. The technique was capable of measuring arsenic with a detection limit of 2.5×10^{17} atoms m^{-2}. Typical concentrations were measured in the range $10^{18}-10^{20}$ atoms m^{-2}. Fukushima et al. (1987) have also used charged particles for carbon, nitrogen and oxygen but it was necessary to use chemical separation after bombardment.

Thermal neutron activation analysis is used to examine the impurities introduced during the production of semiconductor silicon. The technique is well suited to this application because the ^{31}Si produced on activation of the silicon has a short half-life (2.62 h). Verheijke et al. (1989) analyzed whole wafers of silicon, 150 mm in diameter, for 55 elements. The irradiation scheme is shown in Figure 16.5. Special packaging was required to keep the samples free from contamination and the wafers were stacked in a hermetically sealed quartz glass container for irradiation. The samples were irradiated for 3 d in a specially designed irradiation facility with a thermal neutron flux of 4×10^{17} n m^{-2} s^{-1} and allowed to decay for 2 d before unpacking and rinsing with pure water. The wafers were then cut into four. The quarters were stacked to improve measuring geometry and counted for 16 h. After the first measurement the samples were washed with hydrochloric–nitric acid to remove surface contamination and recounted for 16 h. Finally the samples were etched with hydrofluoric–nitric acid and recounted for 21 h to determine the impurities in the surface layer. Typical results for the impurity content in the bulk of the silicon wafers, measured after surface etching, are shown in Figure 16.5.

A similar problem, involving the measurement of very low levels of impurities which may be detrimental to the operation of the device, is

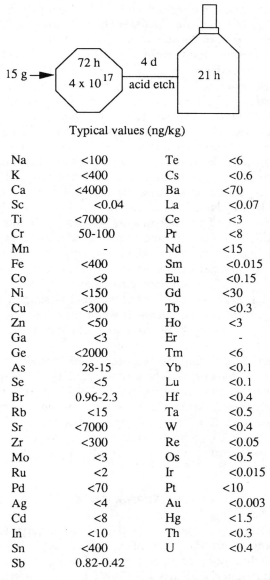

Typical values (ng/kg)

Na	<100	Te	<6
K	<400	Cs	<0.6
Ca	<4000	Ba	<70
Sc	<0.04	La	<0.07
Ti	<7000	Ce	<3
Cr	50-100	Pr	<8
Mn	-	Nd	<15
Fe	<400	Sm	<0.015
Co	<9	Eu	<0.15
Ni	<150	Gd	<30
Cu	<300	Tb	<0.3
Zn	<50	Ho	<3
Ga	<3	Er	-
Ge	<2000	Tm	<6
As	28-15	Yb	<0.1
Se	<5	Lu	<0.1
Br	0.96-2.3	Hf	<0.4
Rb	<15	Ta	<0.5
Sr	<7000	W	<0.4
Zr	<300	Re	<0.05
Mo	<3	Os	<0.5
Ru	<2	Ir	<0.015
Pd	<70	Pt	<10
Ag	<4	Au	<0.003
Cd	<8	Hg	<1.5
In	<10	Th	<0.3
Sn	<400	U	<0.4
Sb	0.82-0.42		

Figure 16.5. Neutron activation analysis of silicon wafers. (Adapted with permission from Verheijke et al., 1989.)

227

the case of fiber optics. Kobayashi and Shigematsu (1987) determined iron, cobalt, nickel and copper in zirconium fluoride (the major component of raw materials for the fluoride glass fiber) using substoichiometric extraction. The detection limits were 10, 0.01, 1 and 0.1 μg kg^{-1} for Fe, Co, Ni and Cu, respectively.

Neutron depth profiling is a technique which can be used to analyze light elements such as helium, lithium, boron and nitrogen (Downing et al., 1987). The sample is exposed to a well-collimated beam of thermal neutrons. While most of the neutrons pass through the sample, reactive atoms will capture a neutron and produce a monoenergetic charged particle. Since the particles travel in straight lines and lose energy as they travel, the depth of the source atom can be calculated from the loss of energy. Typical detection limits for elements such as He, Li, Be, B, N, O, Na, S, Cl, K and Ni are between 10^{17} and 10^{21} atoms m^{-2}. The technique has been used for borosilicate glasses, with the $^{10}B(n,\alpha)^{7}Li$ reaction. The alpha particle emitted is measured with a surface barrier detector and the energy of the particle is used to calculate the depth of the boron. Prompt gamma ray analysis is a sensitive method for boron and it has been used in a routine way to measure boron in borosilicate glass (Riley and Lindstrom, 1987). The method is simple, the samples are crushed and mixed with carbon and 0.5 g of the mixture are pelletized. Samples of glass containing between 1 and 8% boron are analyzed in 15–30 min.

Proton-induced X-ray emission has been used to analyze borosilicate glass for a range of elements (Borbely-Kiss et al., 1985). Major elements Al, Si, K, Ca, and Zn, and trace elements Cl, Ti, Mn, Fe, Cu, As, Rb, Sr and Zr were determined with a 2 MeV proton beam, accelerated in a 5 MV Van de Graaff accelerator, with a beam intensity of 5 nA. Proton-induced gamma emission was also applied to the determination of Na, B, Al and Si using a typical beam intensity of 10 nA.

CLAYS AND CERAMICS

Examples of methods for the activation analysis of clays and ceramics tend to be confined to the field of archaeometry. There are many examples in the literature of fingerprinting for the elucidation of the source of ceramic material (Harbottle, 1976; Perlman, 1981). Clays and ceramics are analyzed in the same way as geological material and the scheme adopted by the Natural History Museum, London is typical (Williams and Wall, 1990). The procedure, which is shown in Figure 16.6, is designed for optimum sensitivity and consists of several very long

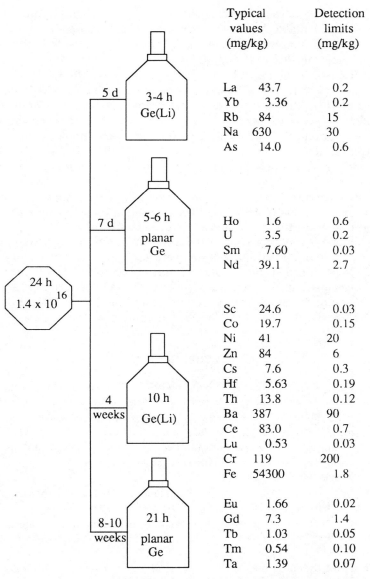

	Typical values (mg/kg)	Detection limits (mg/kg)
La	43.7	0.2
Yb	3.36	0.2
Rb	84	15
Na	630	30
As	14.0	0.6
Ho	1.6	0.6
U	3.5	0.2
Sm	7.60	0.03
Nd	39.1	2.7
Sc	24.6	0.03
Co	19.7	0.15
Ni	41	20
Zn	84	6
Cs	7.6	0.3
Hf	5.63	0.19
Th	13.8	0.12
Ba	387	90
Ce	83.0	0.7
Lu	0.53	0.03
Cr	119	200
Fe	54300	1.8
Eu	1.66	0.02
Gd	7.3	1.4
Tb	1.03	0.05
Tm	0.54	0.10
Ta	1.39	0.07

Figure 16.6. Neutron activation analysis of pottery. (Adapted with permission from Williams and Wall, 1990.)

229

counts after long decay periods. Unlike the rapid results required by industry it is acceptable for archaeological samples to take a little longer. The samples weighing between 50 and 150 mg are irradiated for 24 h in a thermal neutron flux of 1.4×10^{16} n m^{-2} s^{-1}. After 5 days they are counted for 3–4 h for La, Yb, Sb, Rb, Na and As. After 7 d they are recounted for 5–6 h with a planar low energy photon detector for Ho, U, Sm and Nd. After 4 weeks they are recounted on a lithium-drifted germanium detector for 10 h for Sc, Co, Ni, Zn, Cs, Hf, Th, Cr, Ba, Sb, Yb, Lu, Sr and Fe. Finally after 8–10 weeks each sample is measured on the planar detector for 12 h for Eu, Ce, Gd, Tb, Tm and Ta. The results obtained for a pottery sample, plus the detection limits are given in Figure 16.6.

Protons have also been used to analyze pentasil type boroaluminosilicates (Frey et al., 1988). These zeolites are used in industrial applications in adsorption and catalytic processes. Proton beams of 1.7 MeV and 1.3 mm diameter were obtained from the 2 MV Van de Graaff accelerator. Proton currents of 20 nA and 200 nA were used to collect charges of 30 μC and 200 μC for the proton induced X-ray emission and proton induced gamma ray emission analyses, respectively. The lighter elements B, O, Na and Al were measured with gamma rays and the heavier elements Si, S, Cl, K, Ca, Ti, Cr, Mn, Fe, Co, Ni, Cu and Zn with X-rays.

REFERENCES

Aardsma, G.E., P. Jagam, and J.J. Simpson (1987), "Neutron-activation analysis of thorium in acrylic samples," *J. Radioanal. Nucl. Chem.*, **111**(1), 111–116.

Al-Jobori, S.M. (1988), "Determination of impurities in zircaloy clad by means of neutron activation analysis", *J. Radioanal. Nucl. Chem.*, **120**(1), 141–146.

Bellido, L.F. and B. de C. Arezzo (1986), "Non-destructive analysis of inorganic impurities in Brazilian coals by epithermal neutron activation," *J. Radioanal. Nucl. Chem.*, **100**(1), 21–29.

Berti, M. and A. V. Drigo (1982), "Simultaneous nuclear microanalysis of nitrogen and oxygen on silicon," *Nucl. Instr. Meth.*, **201**, 473–479.

Bild, R. W. (1982), "Simultaneous determination of antimony and chlorine in synthetic rubbers by 14 MeV neutron activation analysis," *J. Radioanal. Chem.*, **72**(1–2), 23–33.

Birattari, C., M. Bonardi, and M. C. Cantone (1987), "Determination of arsenic in silicon matrices by proton activation analysis," *J. Radioanal. Nucl. Chem.*, **113**(2), 309–316.

Bode, P. and M. de Bruin (1988), "An automated system for activation analysis with short half-life radionuclides using a carbonfiber irradiation facility," *J. Radioanal. Nucl. Chem.*, **123**(2), 365–375.

Borbely-Kiss, L., M. Jozsa, A. Z. Kiss, E. Koltay, B. Nyako, E. Somorjai, Gy. Szabo, and S. Seif El-Nasr (1985), "Determination of elemental constituents in high voltage insulator borosilicate glasses under proton bombardment", *J. Radioanal. Nucl. Chem.*, **92**(2), 391–398.

Borsaru, M. and P. L. Eisler (1981), "Simultaneous determination of silica and alumina in bulk bauxite samples by fast neutron activation," *Anal. Chem.*, **53**, 1751–1754.

Borsaru, M. and P. J. Mathew (1982), "Fast neutron activation analysis of bulk coal samples for alumina, silica and ash," *Anal. Chim. Acta.*, **142**, 349–354.

Borsaru, M., R. J. Holmes, and P. J. Mathew (1983), "Bulk analysis using nuclear techniques," *Int. J. Appl. Radiat. Isot.*, **34**(1), 397–405.

Burmistenko, U. N., V. A. Glukhikh, and I. N. Ivanov (1977), "Kompleks apparatury dlya fotoyadernogo aktivatsionnogo opredeleniya zolota i soputstvuyushikh elementov v rudnykh probakh," in *Tezisy Dokladov IV Vsesoyuznogo Soveshchaniya po Aktivatsionnomu Analizu*, Tbilisi.

Casper, L. A. (ed.) (1986), *Microelectronics Processing: Inorganic Materials Characterization*, ACS Symposium Series 295, American Chemical Society, Washington.

Chalouhi, Ch., E. Hourani, R. Loos, and S. Melki (1982), "Absolute determination of copper and silver in ancient coins using 14 MeV neutrons," *Nucl. Instr. Meth.*, **200**, 553–560.

Chiba, M. (1981), "Determination of fluorine in steel-making slags by the 14 MeV neutron activation analysis," *Trans. Nat. Res. Inst. Met.*, **23**(4), 265–272.

Clayton, G. C. and C. F. Coleman (1986), "Current developments and applications of nuclear techniques in the coal industry," *Gamma, X-ray and Neutron Techniques in the Coal Industry, Proc. IAEA Advisory Group meeting, Vienna, December 1984*, International Atomic Energy Agency, Vienna, pp. 1–24.

Cunningham, J. B., B. D. Sowerby, P. T. Rafter, and R. Greenwood-Smith (1984), "Bulk analysis of sulphur, lead, zinc, and iron in lead sinter feed using neutron inelastic scatter γ-rays," *Int. J. Appl. Radiat. Isot.*, **35**(7), 635–643.

Dams, R., F. De Corte, J. Hertogen, J. Hoste, W. Maenhaut, and F. Adams (1976), "Activation Analysis – Part 1: Applications to the analysis of high purity materials," in T. S. West (ed.), *Physical Chemistry Series Two, International Review of Science*, Butterworths, London, pp. 40–47.

Davies, J. A., P. A. Hart, and A. C. Jefferies (1986), "Trace multielement analysis in boron materials by epithermal neutron activation analysis," *J. Radioanal. Nucl. Chem.*, **98**(2), 275–287.

Downing, R. G., J. T. Maki, and R. F. Fleming (1986), "Application of neutron depth profiling to microelectronic materials processing," in L. A. Casper (ed.), *Microelectronics Processing: Inorganic Materials Characterization*, ACS Symposium Series 295, American Chemical Society, Washington, pp. 163–180.

Downing, R. G., J. T. Maki, and R. F. Fleming (1987), "Analytical applications of neutron depth profiling," *J. Radioanal. Nucl. Chem.*, **112**(1), 33–46.

Elayi, A. G., P. Damiani, P. Collet, K. Gruel, F. Widemenn, G. Grenier, and D. Parisot (1985), "14 MeV neutron activation analysis of Gaulish silver alloyed coins," *J. Radioanal. Nucl. Chem.*, **89**(2), 113–122.

Fesq, H. W., D. M. Bibby, J. P. F. Sellschop, and J. I. W. Watterson (1973), "The determination of trace-element impurities in natural diamonds by instrumental neutron activation analysis," *J. Radioanal. Chem.*, **17**, 195–216.

Filby, R. H., G. J. Van Berkel, A. E. Bragg, A. Joubert, W. E. Robinson, and C.A. Grimm (1985), "Evaluation of residual fuel oil standard reference materials as trace element standards," *J. Radioanal. Nucl. Chem.*, **91**(2), 361–368.

Frey, H. U., G. Otto, W. Reschetilowski, B. Unger, and K.-P. Wendlandt (1988), "PIGE and PIXE analyses of pentasil type boroaluminosilicates," *J. Radioanal. Nucl. Chem.*, **120**(2), 281–287.

Fukushima, H., T. Kimura, H. Hamaguchi, T. Nozaki, Y. Itoh, and Y. Ohkubo (1987), "Routine determination of light elements by charged-particle activation analysis", *J. Radioanal. Nucl. Chem.*, **112**(2), 415–423.

Germani, M. S., I. Gokmen, A. C. Sigleo, G. S. Kowalczyk, I. Olmez, A. M. Small, D. L. Anderson, M. P. Failey, M. C. Gulovali, C. E. Choquette, E. A. Lepel, G. E. Gordon, and W. H. Zoller (1980), "Concentrations of elements in the National Bureau of Standards' bituminous and subbituminous coal standard reference materials," *Anal. Chem.*, **52**, 240–245.

Guinn, V. P., S. R. Fier, C. L. Heye, and T. H. Jourdan (1987), "New studies in forensic neutron activation analysis," *J. Radioanal. Nucl. Chem.*, **114**(2), 265–273.

Harbottle, G. (1976), "Activation analysis in archaeology," in G. W. A. Newton (ed.), *Radiochemistry, Volume 3*, The Chemical Society, London.

Holmes, R. J., A. J. Messenger, and J. G. Miles (1980), "Dynamic trial of an on-stream analyser for iron ore fines," *Proc. Australas. Inst. Min. Metall.*, **274**, 17–22.

Holtta, P. and R. J. Rosenberg (1987), "Determination of the elemental composition of copper and bronze objects by neutron activation analysis," *J. Radioanal. Nucl. Chem.*, **114**(2), 403–408.

Hoste, J. and C. Vandecasteele (1987), "Determination of light elements in metals and semiconductors by charged particle activation analysis," *J. Radioanal. Nucl. Chem.*, **110**(2), 427–439.

Iskander, F. Y., D. E. Klein, and T.L. Bauer (1986), "Determination of trace element impurities in aspirin tablets by neutron activation analysis," *J. Radioanal. Nucl. Chem.*, **97**(2), 353–357.

Kanias, G. D. (1987a), "Simultaneous determination of chlorine and sodium in large volume parenteral solutions by instrumental neutron activation analysis," *J. Radioanal. Nucl. Chem.*, **116**(2), 347–354.

Kanias, G. D. (1987b), "Instrumental neutron activation analysis of iron and zinc in compact cosmetic products," *Fresenius Z. Anal. Chem.*, **327**, 351–354.

Kanias, G. D. and N. H. Choulis (1985), "Simple and direct determination of active ingredients in drugs by neutron activation analysis," *J. Radioanal. Nucl. Chem.*, **88**(2), 281–291.

Kishi, T. (1987), "Forensic neutron activation analysis. The Japanese scene," *J. Radioanal. Nucl. Chem.*, **114**(2), 275–280.

Klie, J. H. and H. D. Sharma (1982), "Sulfur determination in coal by 14 MeV neutron activation analysis," *J. Radioanal. Chem.*, **71**(1–2), 299–309.

Kobayashi, K. and T. Shigematsu (1987), "Trace determination of iron, cobalt, nickel and copper in zirconium fluoride by substoichiometric radioactivation analysis," *J. Radioanal. Nucl. Chem.*, **113**(2), 333–341.

Kucera, J. and L. Soukal (1983), "Determination of trace element levels in polyethylene by instrumental neutron activation analysis," *J. Radioanal. Chem.*, **80**(1–2), 121–127.

Lindstrom, R. M. (1986), "Activation analysis of electronics materials," in L. A. Casper (ed.), *Microelectronics Processing: Inorganic Materials Characterization*, ACS Symposium Series 295, American Chemical Society, Washington, pp. 294–307.

Lubkowitz, J. A., H. D. Buenafama, and V. A. Ferrari (1980), "Computer controlled system for the automatic neutron activation analysis of vanadium in petroleum with a californium-252 source," *Anal. Chem.*, **52**, 233–239.

Perlman, I. (1981), "Applications to archaeology," in S. Amiel (ed.), *Nondestructive Activation Analysis*, Elsevier, Amsterdam, pp. 259–279.

Pitrus, R. K. (1988), "INAA of trace impurities in aluminium clads," *J. Radioanal. Nucl. Chem.*, **120**(1), 133–139.

Pringle, T. G., S. Landsberger, W. F. Davidson, and R. E. Jervis (1985), "Determination of trace and minor elements in coal: a comparison between instrumental photon and thermal neutron activation analysis," *J. Radioanal. Nucl. Chem.*, **90**(2), 363–372.

Riley Jr, J. E. and R. M. Lindstrom (1987), "Determination of boron in borosilicate glasses by neutron capture prompt gamma-ray activation analysis," *J. Radioanal. Nucl. Chem.*, **109**(1), 109–115.

Rosenberg, R. J. (1985), "Determination of the silver content of ancient silver coins by neutron activation analysis," *J. Radioanal. Nucl. Chem.*, **92**(1), 171–176.

Rousset, M. and M. Federoff (1985), "Multielement instrumental neutron activation analysis of silver–copper coins", *J. Radioanal. Nucl. Chem.*, **92**(1), 159–170.

Sun Jingxin, and R. E. Jervis (1987), "Concentrations and distributions of trace and minor elements in Chinese and Canadian coals and ashes," *J. Radioanal. Nucl. Chem.*, **114**(1), 89–99.

Valkovic, V. (1983), *Trace Elements in Coal, Volume II*, CRC Press, Florida.

Verheijke, M. L., H. J. J. Jaspers, and J. M. G. Hanssen (1989), "Neutron activation analysis of very pure silicon wafers," *J. Radioanal. Nucl. Chem.*, **131**(1), 197–214.

Williams, C. T. and F. Wall (1990), "An INAA scheme for the routine determination of 27 elements in geological and archaeological samples," *British Museum Occasional Publications* (in press).

INDEX

235